PRESENTATION

This unpublished Article is the restructuring of Article 1 "Mathematical Functions of the Periodic Table and the New Quantum-Periodic Mechanics" published in Spanish and English. The following published articles are entitled:

Second article

Solution of the Radial Equation of Hidrogen-Type Atoms and the New Quantum _ Periodic Mechanics.

Third Article

The new quantum-periodic mechanics and the properties of the radial and angular wave function in the design of the periodic table

Fourth Article

Algebra and Geometry of the Periodic Systems of Chemical Elements.

Appeal to the scientific community

I want to note that this longstanding work is haute couture and plus size, but fascinating. It is aimed at science and engineering teachers, students, and scientists around the world. It is a **_new algebra-geometry quantum-periodic_** that not only allows us to calculate the functions of the Periodic Table but, also, its algebra and geometry as well as any other periodic system since the Periodic Table is not unique. Consequently, you will encounter situations that will perplex you but come back again and again to see each deduction and try to see your interpretation in the context of physical reality. I hope it will be refereed, not by two or four or twenty but by the entire scientific community and will be included in the corresponding chair of each university in the world and the New Quantum-Periodic Mechanics will be systematically developed together with Modern Physics and Chemistry.

Abstract

In the present work, only the results for hydrogen-type atoms will be developed from the mathematical solution of the Schrödinger wave equation and from these results how the Periodic Table is mathematically constructed. Discrete period functions are obtained for each block according to the orbitals that are being filled with electrons. This quantization of the atom can only acquire a series of discrete values separated by forbidden intermediate quantities. The discrete values imply the universal Planck constant and these values are modeled by the quantum numbers of energy, momentum and projection plus the so-called intrinsic spin quantum number which, as it will be developed, depends on the sequence of the periodic system to be built. The corresponding functions for each block s, p, d, f, ... allow us to obtain

1) the **ATOMIC NUMBER OF EACH ELEMENT** according to each differential electron [r = 1,2, ...]. This differential electron indicates the column of the block s, p, d, f, ... counted from left to right (or from bottom to top).

2) Functions for the electronic configurations of atoms even those elements of blocks d and f that deviate from the order in the Periodic Table. The functions agree with the experimental results and the Pauli exclusion principle.

3) The function for the length of each period of the Periodic System of Chemical Elements (**PSCE**) which not only results in the number of elements but also the type of subshell that will be filled in that period.

4) Brief introduction to the periodic-quantum study of the H atom that does not affect the mathematical treatment describe of the PSCE but that brings us to the Periodic Table. Articles 2, 3 and 4 elaborate on this topic.

Keywords: Quantum results of hydrogen atoms (sub-shell, orbitals and elements) that lead to Mathematical Functions for each block of elements of the Periodic Table; periodic length; functions for electronic configurations of atoms and PSCE; quantum vacuum of interaction and zero element.

Table of Contents

Introduction ... 7   Methodology ... 8

§ 1 Origin of functions for atomic numbers 11

§ 2 Calculation of the functions for each block s, p, d and f 15

§ 3 Block of elements where the p orbitals are filling 18

§ 4 Block of elements where the d orbitals are filling 21

   § 4.1 Calculation of the function for the elements of the d-block 21

§ 5 Block of elements where the f-orbitals are filling 25

§ 6 Block of elements where the s orbitals are filling 26

§ 7 Functions for electronic configurations 28

   § 7.1 Function for the electronic configuration of the s-block 28

   § 7.2 Function for the electronic configuration of the p-block 31

   § 7.3 Function for the electronic configuration of the d-block 31

   § 7.4 Function for the electronic configuration of the f-block 36

§ 8 Function for length in the Periodic Table 38

   § 8.1 Periodic length function obtained from the ratio that determines the energy levels of hydrogen atoms leading to the Periodic Table ......
   ...43

§ 8.2 Orbital energies and filling for s-block elements 53

§ 8.3 Orbital energies and filling for p-block elements 61

§ 8.4 Orbital energies and filling for d-block elements 69

§ 8.5 Orbital energies and filling for f-block elements 75

§ 9 Function that determines the position of the element from the ns^1 orbital at the beginning of the period and for each column of each block s, p, d, f .....
......82

§ 9.1 Position function for s orbitals 82

§ 9.2 Position function for p orbitals 85

§ 9.3 Position function for d orbitals 86

§ 9.4 Position function for d orbitals 86

§ 10 Results and discussion 87

§ 11 Conclusions and summary of functions and configurations.
…..91

The Electron Configurations of the Elements: Long electronic functions ..
…..95

Appendix … 101

Gratitude 148

Bibliography 148

Introduction

Much has been written in the development of the Periodic Table since the brilliant and lion-looking for his long and untamed hair Dimitri Ivanovich Mendeleev in March 1869 released his Table of chemical elements ordered in cells or grids according to the order increasing their atomic weights and their physical and chemical properties without knowing that 150 years later it became known that these cells are governed by mathematical functions of an independent variable that correspond to the levels of the atom or the period in the Table of Chemical Elements. This first article (chapter) of four focuses on the process involved in obtaining these functions and their relation to the Table cells and going much further. The current explanation of the ordering of the elements has its roots in the quantum mechanics of hydrogen atoms and electronic configurations, it is therefore clear that there are mathematical functions for the configurations as they are in fact presented in this article. These functions are deduced from some "Barrier Diagrams and Electronic Gates", BDEG, which allow visualizing the electronic transit between odd and even states and explain the apparent anomaly in the configuration of some elements such as Cr and Cu. We also find that the periodic length repeats in a consecutive odd and even period and in turn the mathematical function for the length of each period is disclosed ....

Finally, and as a preamble, it is explained how to obtain the periodic length from the relationship that determines the energy levels of the atoms, which relationship connects the new knowledge of periodic algebra with quantum mechanics, hence the origin of the new quantum mechanics -periodic.

In summary: many scientists have contributed new information in the development of the Periodic Table and now we have reached its mathematical functions where organisms such as the IUPAC and Chemical Societies of the world, and specifically the American Chemical Society, ACS, , have allowed the dissemination with greater Academic and Geographical scope of each new advance in the fascinating and cornerstone of Chemistry, Physics, astrophysics, biology and science in general: THE PERIODIC TABLE..

Methodology

Below we will describe the steps to follow to determine the mathematical functions of the Periodic Table:

1) we start from the results of the study of the quantum mechanics of hydrogen atoms (nucleus of Z protons and a single electron) which lead to the structure of atoms in energy levels and sublevels.

2) The n^2 orbitals are distributed in the atoms in increasing order of energy as follows:

$$\left[1s^1\right], \left[2s^1 2p^3\right], \left[3s^1 3p^3 3d^5\right], \left[4s^1 4p^3 4d^5 4f^7\right], \ldots$$ where each

square bracket corresponds to a level or period; also, the number followed by the sublayer inside the bracket represents the energy level. However, this is not the order of filling of the chemical elements;

3) experimentally, the order of filling the orbitals in the sublayers of the atoms has been achieved as follows:

$$\{[1s]\}, \{[2s, 2p]\, [3s, 3p]\}, \{[4s, 3d, 4p]\, [5s, 4d, 5p]\}, \{[6s, 4f, 5d, 6p]\, [7s, 5f, 6d, 7p]\}$$

Here each $\{key\}$ corresponds to a level and each $[bracket]$ to a period. A key $\{\ \}$ indicates the principal quantum number $n = 1,2,3$ y 4 (only 4 keys $\{\ \}$) for the maximum value of the electron impulse $\ell = n - 1$ and the bracket comprises the orbitals of the atoms to be filled in the period that is identified by the corresponding number for the sublayers s and this number (period), in addition, indicates the levels of the atoms of the elements in that period of the Periodic Table. In summary:

at each level ({key}) **the filling of the orbitals of a new sublayer begins**

$l = n - 1$ $\forall l = 0$ (*sublevel* s); $l = 1$ (p); $l = 2$ (d) and $l = 3$ (f) up to n = 4;

4) each orbital is allowed two electrons and the chemical elements result. Now we model the previous sequence that leads to the periodic system in figure 1.1 [important note: to understand the quantum-periodic mechanics it takes a few figures to model the phenomenon and then carry out the analytical calculations and compare the results with each figure. The corresponding calculations are explained in the following articles (chapters)].

5) We transform figure 1.1 into a **staggered function** at the corresponding periodic interval being the elements that have filled the orbitals or the p states in the upper part of the X axis and under the staggered function n = **1, 2, 3 y 4 as** modeled in figure 1.2. Now we turn to the differential and integral calculation but adjusted to a real physical phenomenon so that the area to be determined fits with this phenomenon, for which the integration is carried out by adding the states according to how the system was sequenced, a calculation explained in the corresponding section;

6) Then a series of displacements is made to the levels of the system in figure 1.2 so that the rest of the states remain as areas and the function corresponding to each state can be determined, a result that appears in figure 1.4. In each case, to find the function, the integration starts at zero

giving the expected results of the real phenomenon and which will be further explored;

7) The above calculation applies to the elements that have completely filled the orbitals of the sub-layer obtaining the corresponding atomic numbers according to the number of levels of the atom which corresponds to the periodic variable x of the functions for each state s, p, d and f. Then the function for the atomic numbers at the beginning of each block is determined;

8) the previous functions allow us to calculate the atomic numbers and their electronic configuration without difficulty for the elements of the s and p blocks; however, for the elements of blocks d and f it is necessary to refer to some **"Barrier Diagram and Electric Gates, BDEG"** that indicate how to solve certain irregularities in the configurations and allow us to obtain functions for the configurations of these elements;

9) the function for the elements that have filled the s orbitals allows us to determine another function for the length of each period in the table that gives us, not only the number of elements in the period but which orbitals the elements will use in that period; here it is necessary to start from scratch for physical reality to exist;

10) finally, and as a preamble we connect a quantum result with periodic analysis; that is, **the relationship that determines the energy levels of hydrogen atoms** with the selected sequence that will lead to the periodic table or any other system which will be studied in detail later. Articles 2, 3 and 4 elaborate on this topic.

§1. Origin of functions for atomic numbers

When solving the equation Schroedinger[1] applied to the atoms with Z protons and a single electron, the following results are obtained:

A) for each value of n there are n possible values for *l*. That is, _each layer (shell) n has n sublayers (subshells)._ If, for example, n = 3 represents the M **shell** with three sublayers: 3s, 3p and 3d for, respectively. Although **"each shell n has n subshells"** quantum mechanics does **NOT** establish the order in which these sublayers are going to be filled; the periodic algebra yes. The sequence or order of filling of the orbitals is reflected in figure 1.1.

B) For each value of *l* $(2l+1)$ is possible values for **m** that goes from $-l$ a $+l$ passing through zero. That is, each subshell is divided into orbitals. For example, subshell p has 3 values for m = -1, 0, 1.

C) For each value of $n \equiv n$ there is a total of n^2 ORBITALS, it is true that

$$\sum_{l=0}^{n-1}(2l+1) = 1 + 3 + 5 + 7 + \ldots = n^2$$

D) Experimentally [2] the order of filling of the subshells for atoms with Z electrons has been determined. In general, all atoms produce a single spectrum such as a fingerprint. Although in an atom of several electrons the striped spectrum is more complex than the spectrum of the hydrogen atom these spectra can be interpreted based on the same results found in quantum mechanics for the hydrogen atom (A, B and C). By analyzing the spectrum of all elements and determining their quantum numbers, n, ℓ the Periodic Table has been obtained. In the spectroscopic study, X-ray and other techniques of all the elements it has been found that, except for some elements that will be explained based on periodic algebra, the filling sequence of the subshells is as follows (a zero period (level) has been included without which the quantum-periodic results would not make physical sense):

$$\{[0s],[1s]\};\ \{[2s,2p],[3s,3p]\};\ \{[4s,3d,4p],[5s,4d,5p]\};\ \{[6s,4f,5d,6p],[7s,5f,6d,7p]\}$$

(1.1)

Here each key $\{\ \}$ corresponds to a level and each bracket $[\ ]$ to a period. Figure 1.1 illustrates the fill sequence for the SPEQ and the above nomenclature for braces and brackets. For example, below level n = 3, periods 4 and 5 appear and physically means that the orbitals corresponding to these two periods are going to be filled with electrons with the maximum angular momentum $l = 2$ characteristic of the orbitals or subshell d, despite using levels 4 and 5 of the atoms.

From this order or filling sequence, figure 1.1, the calculation of the functions for each block of elements begins, treating the atomic numbers as areas. Note that at the end of each level (here only four, n = 4, infinity) the orbitals of the subshells for the electronic states $l = 1$ (p states) will be filled.

The order of filling of the atomic subshells represented in figure 1.1 is modeled in a _rectangular coordinate system_ resulting in a **_stepped function_** (**_staggered function_**), figure 1.2: a function for any of the blocks s, p, d and f of the PSCE defined for $0 < x < 7$ is divided into the following intervals (the values $(\ ),[\ ]$ that the function takes at the ends of the subintervals does not matter):

$$n=1\ (l=0):\ (0,1);\ n=2\ (l=1):\ (1,3);\ n=3\ (l=2):\ (3,5);\ n=4\ (l=3):\ (5,7) \quad \textbf{(1.2)}$$

The intervals indicated in (1.2) for figure 1.2 (left) correspond, respectively, with the even and odd periods $[x_0, x_f]$ defined by $x_0 = 2(n-1)$ y $x_f = 2n - 1$ [see relation (1.1)] for all $n \geq 1$:

$n = 1$ $(l=0)$: $[0,1]$; $n = 2$ $(l=1)$: $[2,3]$; $n = 3$ $(l=2)$: $[4,5]$; $n = 4$ $(l=3)$: $[6,7]$

(1.3)

Consequently, **the length of each level is equal to two** (**stepped function**) covering a consecutive odd and even period. In addition, we observed in the PSCE of figure 1.2 left [this would be another way of representing the PSCE or Periodic Table and makes possible the mathematical solution of it] that for the **p states** (Block p: Bp), colored in green, the area under the *stepped function* corresponds to 2, 10, 18, 36, 54, 86 and 118 in the periods, respectively, $x = 1, 2, 3, 4, 5, 6$ y 7 which comprises both x_0 como x_f. Note that the height between the zero level (X axis) and the level n = 1 is 2; the height between n = 1 and n = 2 is 6; between n = 2 and n = 3 is 10; between n = 3 and n = 4 is 14 that correspond to the capacity of each subshell s, p, d and f, respectively. The sum of the subshells that gives us the capacity of each level **n** is given by

$$\sum_{l=0}^{n-1} 2(2l+1) = 2 + 6 + 10 + 14 + \ldots = 2n^2 \qquad (1.4)$$

The periodic system to the right of figure1.2 is rotated 90 ° clockwise and we obtain the Periodic Table as shown below (figure1.3):

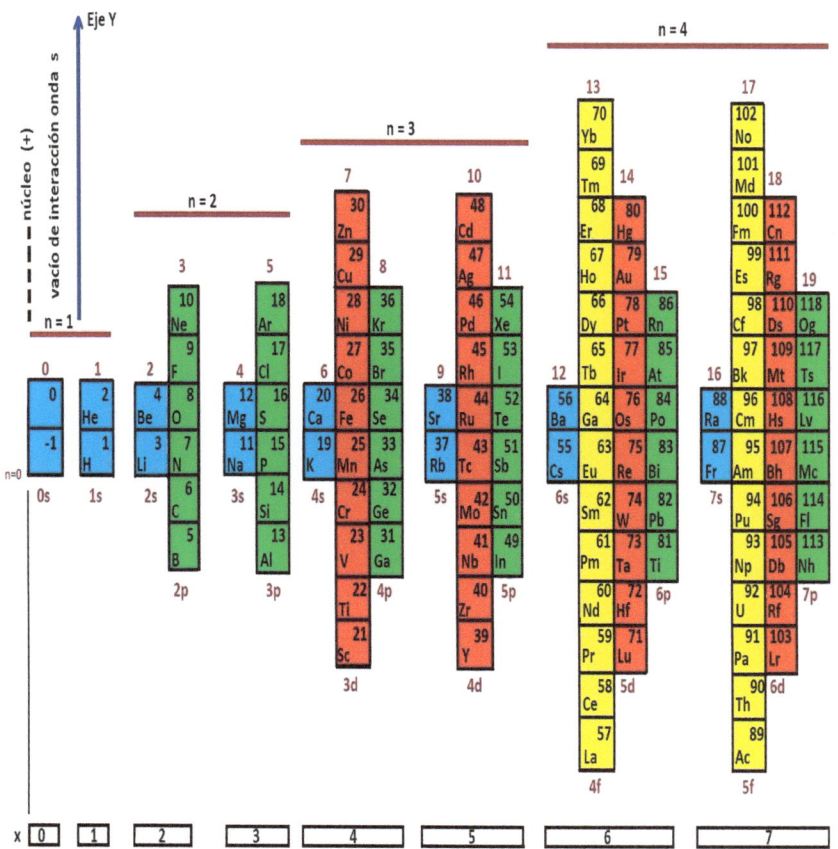

Figure 1.1 This is the filling sequence of the sub-shell orbitals that results in the SPEQ. The allowed values for the angular momentum quantum number l range from zero to n - 1, that is to say, $l = n - 1$. Here we find only four levels at the top (n = 4) which implies that the elements of the Periodic Table will only use electrons up to the maximum $\ell = 3$ corresponding to the states f. Each period begins with the ns subshell so that the period coincides with the level n = x and determines the number of energy levels n = 1, 2, 3, ... that an atom has. The system uses 19 subshells that appear above 7p. The atomic numbers go in consecutive order from bottom to top. This order supports mathematical solution as explained below. To fix an idea: the filling sequence of the p subshells corresponds to the sequence 3, 5, 8, 11, 15, 19, ... and it is about finding the mathematical function for this order; in addition, the function for orbitals and atomic numbers. Note that the **top n** does not correspond to that $n \equiv x$ for the periods. The color agreement is as follows: **Blue**: states s; **Green**: p states; **Red**: states d; **Yellow**: states f.

§ 2 Calculation of the functions for each block s, p, d, f

Now we proceed to the calculation of the functions starting with the block p (B-p) because in each period $x = 1, 2, 3, 4, 5, 6$ y 7 the cycle ends, respectively, with the atomic numbers 2, 10, 18, 36, 54, 86 and 118 that will correspond to our area under the stepped function. We are going to refer to figure 1.2 (left) to calculate the functions that these atomic numbers will give us; however, the calculation can also be done with figure 1.3. First, the calculation is made for the orbitals of the completely filled subshell whose atomic numbers are above under each level (shell) n = 1, 2, 3 and 4. Periodic systems are built in three basic stages: subshell, orbitals, and two electrons per orbital as elements are constructed in equal numbers of protons and electrons in neutral atoms (see figure 1.4). Each of these components or stages is identified with its function for each block. The period zero is identified with the numbers 0 and -1 because when taking differences, it must give the height of the level equal to two in that period: (see figure 1.3) r = 2: 2(He) – 0 = 2 y r=1: 1(H)-(-1) = 2 $s_{r=1,2}(0)$ is calculated as follows: (see calculation of s orbitals and § 8)

$$s(x)_{r=1,2} = [2(x-1)+r] + 6(x-2) + 10(x-4) + 14(x-6) + \ldots \quad x \geq 0$$

And for x = 0:

$$s_{r=1,2}(x) = [2(x-1)+r] \; si \; x=0 \Rightarrow s_{r=1,2}(0) = [2(0-1)+r] = \begin{cases} -2+1=-1 & si \; r=1 \\ -2+2=0 & si \; r=2 \end{cases}$$

r is for the first and second column at level or period zero (x = 0) (figure 1.3)

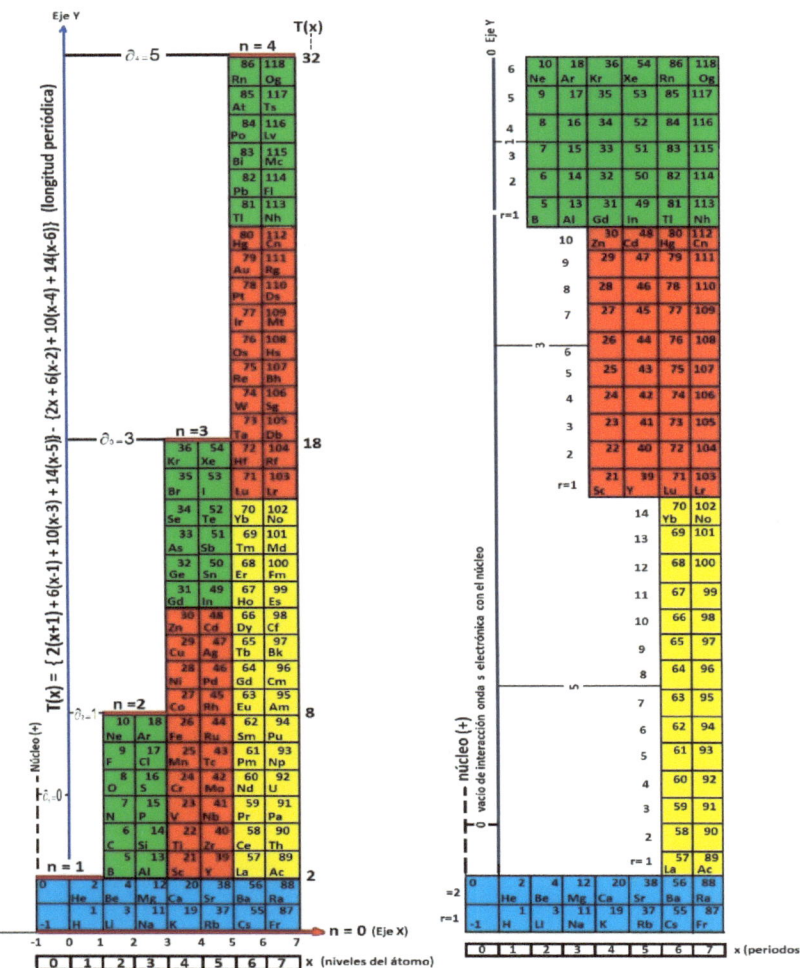

Figure 1.2 From the sequence in Figure 1.1 that leads to PSCE we take it to a coordinate axis system (here left). Note the order in which the orbitals of the subshells are placed to obtain a **stepped function of discrete periods**. On the right, the common blocks of subshells **s, p, d, and f** have been grouped. Two electrons are assigned to each orbital (exclusion principle). The atomic numbers of the elements are treated as areas. Note that the lower limits of integration at the beginning of each level n = 1,2,3,4, ... on the X axis correspond **to the periods before the end of the level** $(n-1)$, that is, if the level n = 2 has the maximum for the angular momentum $\ell = 1$ the lower limit of integration is that for $\ell = 0$ (states s) of level n = 2-1 = 1. **The variable x (periods) is taken as the upper limit**. Below the level n = 1 is the

period n = 0: $(n-1) = (1-1) = 0$ which is the period prior to the start of level n = 1 (see below for an argument in § 4).

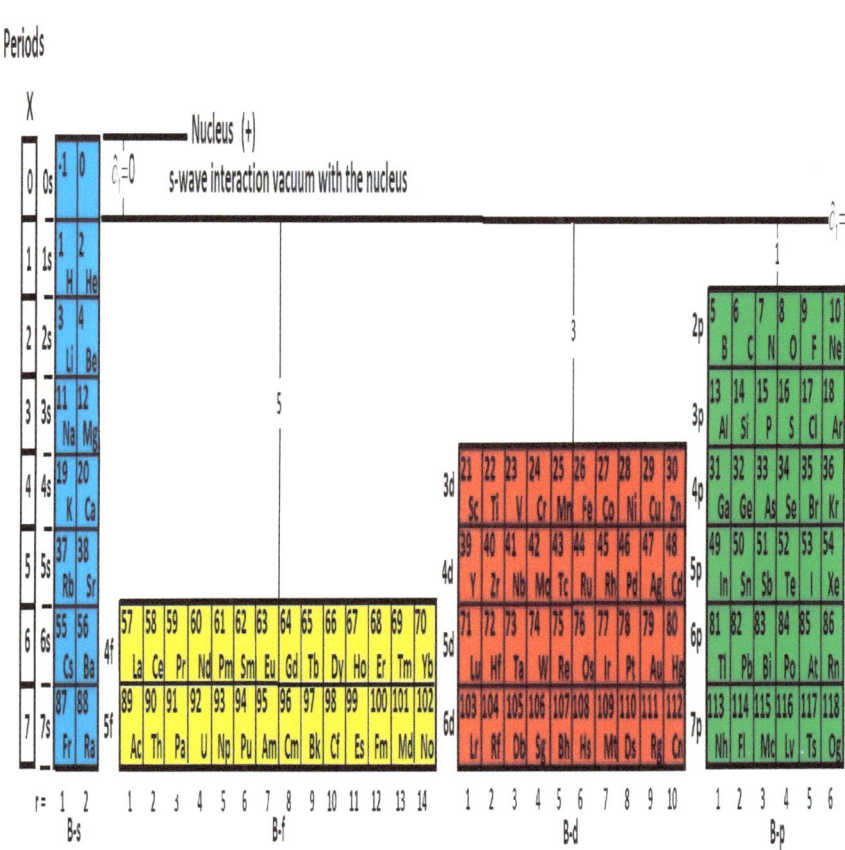

Periodic lenght function: $T(x) = \{2(x+1) + 6(x-1) + 10(x-3) + 14(x-5)\} - \{2x + 6(x-2) + 10(x-4) + 14(x-6)\}$

Orbitals: s   p   d   f   s   p   d   f

Figure 1.3 Notice how we have arrived at the periodic table that we represent here. Atomic numbers are counted from left to right: ***horizontal periods***. In the Periodic Table to the left of figure1.2 the atomic numbers are counted from bottom to top being the ***vertical periods***. The variable $\boxed{r \equiv r}$, below, indicates the columns of each block counted from left to right. Each block will fill the orbitals with angular momentum electrons:

$$l = 0 \, (B-s); \quad l = 1 \, (B-p); \quad l = 2 \, (B-d); \quad l = 3 \, (B-d)$$

§ 3 Block of elements where the orbitals p is filling

The piecewise stepped function for the PSCE (figure 1.2) corresponds to the following lower integration limits, which are the periods before the beginning of each level: $2l \Rightarrow l = 0$ y 1, 3, 5, ..., $(2l+1)$ para $l = 1, 2, 3, ..., n - 1$. (It will already be explained for $2l \Rightarrow l = 0$). The capacity of each subshell $2(2l+1)$ must be integrated and the layer x must be taken as the upper limit of integration. This results in the area under the graph of the **stepped function** whose area corresponds to the atomic numbers 2, 10, 18, 36, 54, 86 and 118. As the maximum value of the subshell p $(l = 1)$ in any layer x = 2, 3, 4, 5, . . . are 6 electrons we get, then, the completely filled subshell: $(x \geq 2)$

$$[Z]p_6(x) \equiv p_6(x) = \sum_{l=0}^{0} \int_{2l}^{x} 2(2l+1)\,dt + \sum_{l=1}^{n-1} \int_{2l-1}^{x} 2(2l+1)\,dt$$

(1.5)

Read: "atomic number Z of the element of block p in period x of column r = 6"

The integration is made as we insert blocks between the base block, B-s, and the periodic block, B-p. Solving the integral (1.5) we obtain:

$$p_6(x) = \sum_{l=0}^{0} 2(2l+1)[x - 2l] + \sum_{l=1}^{n-1} 2(2l+1)[x + 2l - 1)]$$

(1.6)

We develop (1.6) up to n = 4 (only 4 blocks for the PSCE: s, p, d and f) for fully filled subshells results: [The top n in figure 1.1 for 4 blocks: s, p, d and f corresponds to the maximum electronic angular momentum in the filling of orbitals. In order not to confuse this **n of the stepped function** with the **n of the maximum levels** we use x for periods or the maximum level occupied by the electrons in the atom]

$$p_6(x) = 2x + 6(x - 1) + 10(x - 3) + 14(x - 5) + \ldots$$
states: s   p   d   f

(1.7)

Note that in periodic functions each term corresponds to a state as reflected in function (1.7). To determine the function by increasing the atomic number by one and adding an electron, we insert the term $\boxed{-[6-r]}$ for the state p into function (1.7) [for each state, the corresponding term is inserted]; r = 1, 2, 3, 4, 5, 6 rows (figure 1.2) or columns for figure 1.3.

$$p(x)_{r=1,2,\ldots,6} = 2x + 6(x - 1) - [6 - r] + 10(x - 3) + 14(x - 5)$$

Remains

$$p(x)_{r=1,2,\ldots,6} = 2x + [6(x - 2) + r] + 10(x - 3) + 14(x - 5)$$

(1.8)

Being r = 1, 2, 3, 4, 5 and 6 differential electrons that occupy the p orbitals in the different subshells 2p, 3p, 4p, ... with two electrons "assigned" by atomic orbital.

Example 1.1

With the function (1.8) determine the atomic number, $Z = p(x)_r$, and its configuration of the element of 7 shells (x = 7) that has entered the orbital p the differential electron r = 3.

Answer

We substitute in (1.8) the data that are: period x = 7 and column r = 3 of the blocks that is filling the p, B-p orbitals:

$$p(x)_{r=3} = 2x + [6(x-2)+r] + 10(x-3) + 14(x-5)$$
$$p(7)_{r=3} = 2(7) + [6(7-2)+3] + 10(7-3) + 14(7-5) = 115$$
$$p(7)_{r=3} = 2(7) + [6(5)+3] + 10(4) + 14(2)$$

CE :
$1s^2$
$2s^2 \quad 2p^6$
$3s^2 \quad 3p^6$
$4s^2 \quad 4p^6 \quad 3d^{10}$
$5s^2 \quad 5p^6 \quad 4d^{10}$
$6s^2 \quad 6p^6 \quad 5d^{10} \quad 4f^{14}$
$7s^2 \quad 7p^3 \quad 6d^{10} \quad 5f^{14}$

Parentheses indicate the number of subshells s, p, d, and f of the element and r = 3 a sub layer p more incomplete. The electronic configuration, CE, written for the filled sequence is as follows:

$$1s^2 \, 2s^2 \, 2p^6 \, 3s^2 \, 3p^6 \, 4s^2 \, 3d^{10} \, 4p^6 \, 5s^2 \, 4d^{10} \, 5p^6 \, 6s^2 \, 4f^{14} \, 5d^{10} \, 6p^6 \, 7s^2 \, 5f^{14} \, 6d^{10} \, 7p^3$$

And for the final filling in order of subshells s, p, d and f and layers n = 1, 2, 3, ..., 7 would be:

$$1s^2 \, 2s^2 \, 2p^6 \, 3s^2 \, 3p^6 \, 3d^{10} \, 4s^2 \, 4p^6 \, 4d^{10} \, 4f^{14} \, 5s^2 \, 5p^6 \, 5d^{10} \, 5f^{14} \, 6s^2 \, 6p^6 \, 6d^{10} \, 7s^2 \, 7p^3$$

§ 4   Block of elements where the orbitals d is being filled

The process now consists of moving to the right the levels of the stepped function of the PSCE, figura 1.2, so that in levels (shells) 3 and 4 and periods x = 4, 5, 6,7, corresponding to subshells d, they remain top the atomic numbers of the elements that are filling the orbitals d and calculate them as areas. **All blocks p, d, f, ... of elements of the Periodic Table have a part in common with the s orbitals (blue color).** With this we get the lower limits of integration. In the following diagram, figure1.4, the displacement has been made to obtain the lower limits of integration that makes it possible to calculate the area function not only of the B-d but of the blocks B-s and B-f.

§ 4.1 Calculation of the function for the elements of full subshells of the block d

In figure 1.4 for the B-d we have the lower limits of integration 0, 2, 3 and 5 (only up to n = 4 blocks of infinity) and the integrating the capacity of each subshells s, p, d y f it is 2, 6, 10, 14, respectively. As upper limit the periodic variable x. The general integration is as follows for the elements that have completely filled the d orbitals, that is, when the differential electron r = 10 has entered the orbital. Notice how the lower limit **2*l*** begins to appear. We have, then

$$d_{10}(x) = \sum_{l=0}^{1} \int_{2l}^{x} 2(2l+1)\,dt + \sum_{l=2}^{n-1} \int_{2l-1}^{x} 2(2l+1)\,dt$$

$$d_{10}(x) = \int_{0}^{x} 2\,dt + \int_{2}^{x} 6\,dt + \int_{3}^{x} 10\,dt + \int_{5}^{x} 14\,dt + \ldots$$

Finally solving the integrals until n = 4 we obtain *the function for the fully filled subshells d* from the periods $x \geq 4$ that the 3d subshell begins to fill:

$$d_{10}(x) = 2x + 6(x-2) + 10(x-3) + 14(x-5) + ...$$
(1.9)

To find the atomic numbers of the elements in the periods $x = 4,5,6,7$ when the differential electrons have entered the orbital d [it will already be explained in the electronic configurations when the electron passes to another orbital but the result for the atomic number is not altered] r = 1, 2, 3, 4, 5, 6, 7, 8, 9, 10 that correspond to the columns counted from left to right in Figure 1.3. In function (1.9) we insert the term $\boxed{-[10-r]}$ in term 10 (x-3) for states d and proceed algebraically, resulting in:

$$d(x)_{r=1,2,...,10} = 2x + 6(x-2) + [10(x-4)+r] + 14(x-5) + ... \quad \forall x \geq 4$$
(1.10)

Read: $[Z]d_r(x) \equiv d(x)_r$ "atomic number Z of the element of block d in column r = 1,2,3,4,5,6,7,8,9,10 of period x (or in period x of column r)

Note: the function for orbitals and subshells, respectively, that would completely fill the elements of block d, B-d, would obviously be: (compare 1.9; see figure 1.1)

$$d(x)_{orbitales} = x + 3\ x - 2) + 5\ x - 3) + 7\ x - 5) \quad \forall x \geq 4$$
$$d(x)_{sucapas} = x + (x-2) + (x-3) + (x-5) \quad \forall x \geq 4$$

(See Figure 1.1 and Figure 1.4).

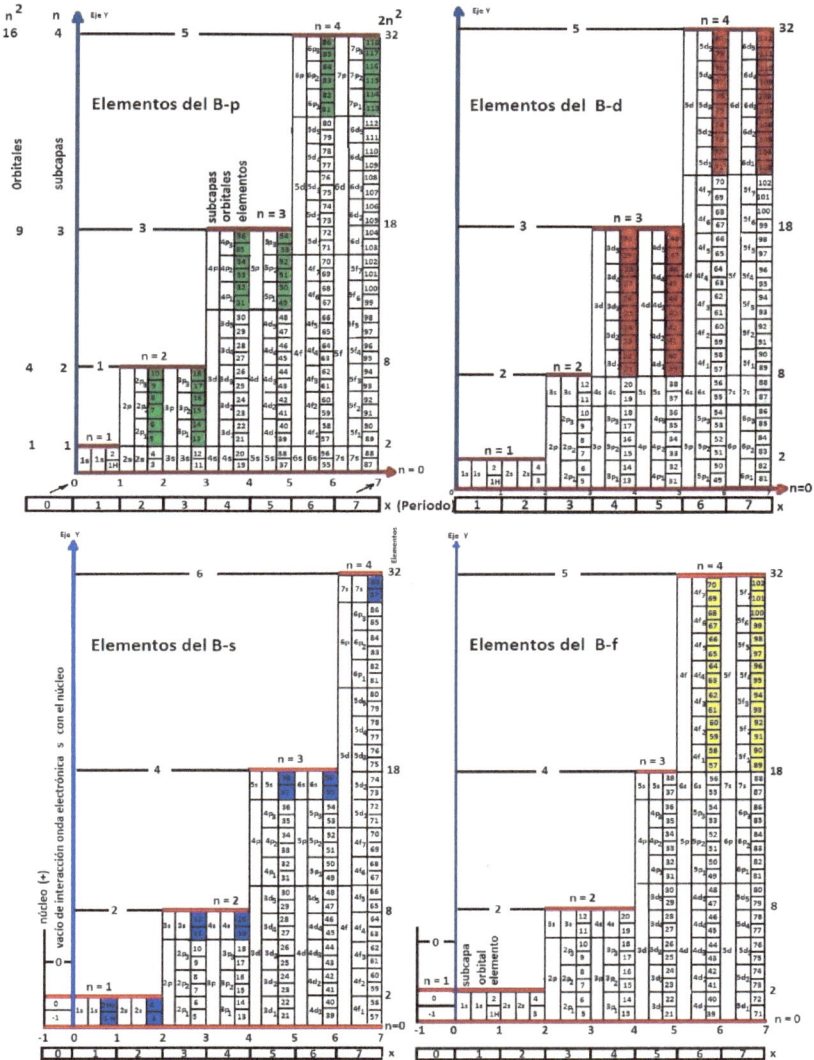

Figure 1.4 Here we have represented the structure of an atom in sublayers, orbitals and two electrons per orbital. The functions calculate Z for the elements of each block and the number and type of orbitals in the subshells for their construction. Each period begins in a subshell s and ends the cycle in subshell p leaving the elements of completely filled layers above. Then the levels (shells) of B-p are shifted as indicated here and the periods prior to the beginning of each level (shell) are taken as lower limits of integration. The nucleus is located in the abscissa -1 due to the quadratic function $(x+1)^2/4$ obtained from the relationship that determines the levels of the atom deduced from the quantum mechanical-quantum study of the hydrogen atom (see § 4). Note: subshell is subcapa; orbital is orbital and elementos is elements.

Example 1.2

> With the function (1.10) determine the atomic number, and its configuration of the element of 5 shells (x = 5) that has entered the orbital d differential electron r = 4 (column 4).

Answer

We substitute in (1.10) the data that are: period x = 5 and column r = 8 of the blocks that is filling the orbitals d, Bd:

$$d(x)_r = 2x + 6(x-2) + [10(x-4)+r] + 14(x-5)+... \quad \forall x \geq 4$$
$$d(5)_{r=4} = 2(5) + 6(5-2) + [10(5-4)+4] + 14(5-5)$$
$$d(5)_{r=4} = 2(5) + 6(3) + [10(1)+4] + 14(0) = 42 \ (Mo)$$

CE: $1s^2$
     $2s^2$ $2p^6$
     $3s^2$ $3p^6$
     $4s^2$ $4p^6$ $3d^{10}$
     $5s^2$ $4d^4$

CE attempt: $1s^2 2s^2 2p^6 3s^2 3p^6 4s^2 3d^{10} 4p^6 5s^2 4d^4$

==

It is clear that the periodic function $Z = d(x)_r$, B-d, correctly locates the element Mo in the block, period and column in the Periodic Table. However, experimentally its configuration is as follows: CE correcta: $1s^2 2s^2 2p^6 3s^2 3p^6 4s^2 3d^{10} 4p^6 5s^1 4d^5$ but its atomic number and its position in the Table remain valid. This configuration is explained through BDED diagrams (Barrier Diagram and Electronic Doors) that show where electronic traffic can occur. For the moment, we present a

BDED for the elements Z = 41 and 42, figure 1.5. This type of diagram will be analyzed later.

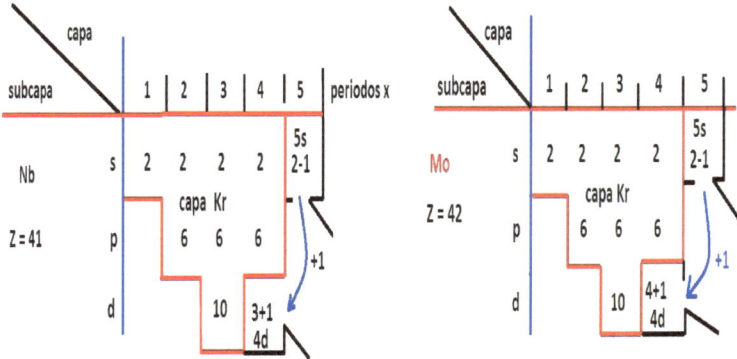

Figure 1.5 Electronic transitions occur between **_odd and even states_** [3] as shown here. In addition, the exclusion principle restricts two electrons per orbital. The 5s orbital does not admit another electron, so the transit occurs from 5s to 4d. Note the electronic doors. The Barrier consists of the Kr layer in a continuous line, hence the abbreviation BDED.

§ 5   Block of elements where the orbitals f is filling

The reasoning described for blocks B-p and B-d is analogous for blocks B-f and B-s. The calculation is made taking the lower limits of integration shown in figure 1.4 and Theas **_upper limit_ the number of levels of the atom** that corresponds to the periodic variable, that is period x. Below is the general integration for n levels (the **n** of the stepped _**function**_ for **n** = 1, 2, 3 and 4 blocks) which corresponds to the completely filled subshells and then as the atomic number increases one by one:

$$f_{14}(x) = \sum_{l=0}^{2} \int_{2l}^{x} 2(2l+1)\,dt + \sum_{l=3}^{n-1} \int_{2l-1}^{x} 2(2l+1)\,dt$$

(1.11)

For completely *filled subshells f* and four blocks (up to n = 4) we obtain:

$$f_{14}(x) = 2x + 6(x-2) + 10(x-4) + 14(x-5)$$

(1.12)

To calculate the atomic numbers of the elements of the block f (n = 4) when they have entered the orbitals f, and fill the different subshells of the corresponding x periods, the differential electrons r = 1, 2, 3, 4, 5, 6, 7, 8, 9, 10, 11, 12, 13 and 14 algebraically rearrange the solution as it was done with the p and d block, inserting $-[14-r]$ it as follows:

$$f(x)_r = 2(x-0) + 6(x-2) + 10(x-4) + [14(x-6)+r], \quad \forall x \geq 6$$

(1.13)

The nomenclature is identical to blocks p and d. The calculation for some element of B-f is done as explained in example 1.1 and 1.2.

§ 6 Block of elements where the orbitals s is filling

In the three previous blocks p, d and f a GENERAL LAW is observed [This Law exists with many properties and is applicable to any periodic system but here the thing of greatest interest is the PERIODIC TABLE as it is the real phenomenon]. for calculating periodic functions. We have started from B-p and shifted the levels in order to calculate the atomic numbers of filled sublayers as areas. Note the lower limits of integration that are determined as the distance of the Y axis at the beginning of each level. In figure 1.4 we see that the limits for the B-s correspond to 0, 2, 4, 6, ... that comply with a formation law $x = 2l \quad \forall l = 0,1,2,3,...,n-1$.[Hence, as previously noted, the term **2l** and how it appears in the general integrations for the lower limits] where $l = 0,1,2,3,...$ is the angular momentum quantum number corresponding to the type of electron that will occupy the levels n (figure 1.1). in the corresponding periods in the construction of the PERIODIC TABLE. In the case of period zero the phenomenon deals with the interaction of the electronic wave s and the nucleus. Following is the calculation for the ***base block function s, B-s:***

$$s_2(x) = \sum_{l=0}^{n-1} \int_{2l}^{x} 2(2l+1)dt = \sum_{l=0}^{n-1} 2(2l+1)[x-2l] = 2x + 6(x-2) + ...$$

(1.14)

The subscript 2, for the function $s_2(x)$ that gives the atomic numbers of the elements that have filled the s orbitals, refers to the column $r = 2$ in figure 1.3.

The function (1.14) calculates the atomic numbers 2, 4, 12, 20, 38, 56, 88, 120, ... in the periods, respectively, $x = 1, 2, 3, 4, 5, 6, 7$ y 8 ... The general function that includes the atomic numbers of the elements H, Li, Na, K, Rb, Cs, Fr, . . . When filling one by one, it is as follows:

$$s(x)_{r=1,2} = \left[2(x-1)+r\right]+6(x-2)+10(x-4)+14(x-6)+\ldots \quad x \geq 0$$

(1.15)

Note that in the systems represented in figures 1.1; 1.2 and 1.3 appear for $x = 0$ the values -1 y 0. Precisely these values are calculated with (1.15) with $x = 0$ y $r = 1,2$

§ 7. Functions for Electronic Configurations

The periodic functions for the Periodic Table and each block are summarized in the Sheet of the new Table on the back and at the end of this manual. Functions are deduced for configurations through Diagrams, as you will see, and noting that the elements where there is electronic transit are in the same period; in addition, according to theoretical results electronic transitions only occur between even and odd states[3]. The functions thus obtained fit with the experimental results.

§ 7.1 Function for the Electronic Configuration of the Elements that fill the orbitals s: Block-s

For the configurations of the elements of blocks B-s and B-p are determined without difficulty with functions (1.8) and (1.15) since there is <u>no</u> exception. There is a large difference in very large energy [2] between each subshell s and the p subshell that precedes it. However, we will give the functions for these blocks that are derived from the BDED without further details to focus on the functions for the B-d and B-fblocks. The configurations of the atoms can be simplified since the chemical elements are characterized by having completely filled shells and subshells in the period before the maximum level where the orbitals are being filled, that is, the level (n - 1). In other

maximum level where the orbitals are being filled, that is, the level (n - 1). In other words, the elements have a noble gas armor. If we start from (1.7) and call $\left[p(x-1)\right]$ the function for internal electrons or noble gas armor shells and subshells completely filled up a previous period, $(x-1)$, to the atomic number element Z of the block s, then (1.7) would be transformed as follows:

$$\left[p(x-1)\right] = 2(x-1) + \sum_{l=1}^{n-1} 2(2l+1)\left[x-2l\right]$$

$$\left[p(x-1)\right] = 2(x-1) + 6(x-2) + 10(x-4) + 14(x-6) + \ldots$$

(1.16)

Example 1.3

Apply the function (1.15) to the elements in the period x = 7 that are filling the s orbitals with r = 1.2 electrons

$$s_{r=1,2}(x) = \left[2(x-1)+r\right] + 6(x-2) + 10(x-4) + 14(x-6): x=7 \text{ y } r=1,2:$$

$$\left[2(6)+1,2\right]s \quad 6p(5) \quad 10d(3) \quad 14f(1) = 86 + r = Z$$

As we can see we find Z = 86 that corresponds to the element Rn (radon).

The configuration of the elements for $r = 1, 2 \Rightarrow r = Z - [p(x-1)]$ are Z = 87 and 88, respectively, for which there is no anomaly, we represent it in the following BDED-s (figure 1.6) From this diagram it is deduced for the configurations by the following **_electronic configuration function_** for the block (s):

$$CE[Z]_s(x)_r = [p(x-1)] \times s^{Z - [2(x-1) + 6(x-2) + 10(x-4) + 14(x-6) + \ldots]}$$

(1.17)

And Z is given by (1.15).

$CE[Z]_s(x)_r$ Read: electronic configuration, CE, of the atomic number element Z of B-s located in period x and column r.

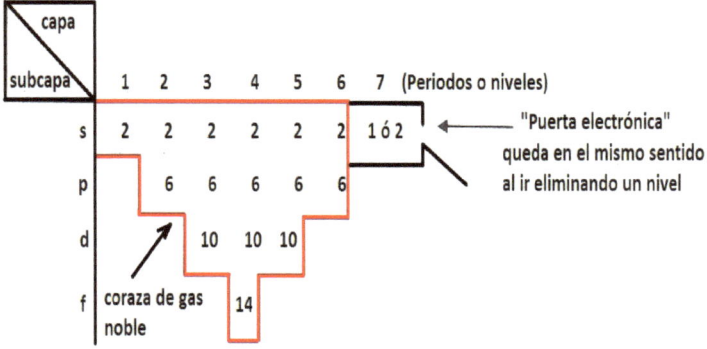

Figure 1.6 Through these diagrams and without calculations of quantum mechanics (theory of variation, perturbation, Hartree-Fock, etc.) it is observed that electrons enter through certain "gates". We note that as one shell is removed, the "electronic door" opens to house 1 or 2 electrons and the previous period (x-1) will always correspond to a noble gas element (here Rn: Z = 86) of fully filled shells and subshells whose electronic distribution is expressed within the continuous line. Note the order in which the electronic configuration is written in shells and subshells.

§ 7.2 Function for the Electronic Configuration of the Elements that fill the orbitals p: Block-p

A reasoning similar to B-s leads us to the following function for B-p:

$$CE[Z]_{p_r}(x) = [p(x-1)] \cdot (x-2)f^{14} \cdot (x-1)d^{10} \cdot x\,s^2 \cdot x\,p^{Z-d_{10}(x)}$$

(1.18)

As the first subshell f is the 4f it follows that $x \geq 6$ in the term $(x-2)f^{14}$. The first subshell d is 3d accordingly $x \geq 4$. The atomic number of the B-p element in period x and column r is given by (1.8) that we rewrite:

$$Z = p(x)_{r=1,2,\ldots,6} = 2x + [6(x-2)+r] + 10(x-3) + 14(x-5)$$

The term $x\,p^{Z-d_{10}(x)}$ is derived from this function just as it was done for the B-s. For $x\,s^2$ here the x corresponds to the number of levels (shells) of the atom.

§ 7.3 Function for the Electronic Configuration of the Elements that fill the orbitals d: Block-d

We start with the function (1.10) that we copy below:

$$Z = d(x)_r = 2x + 6(x-2) + [10(x-4)+r] + 14(x-5)+\ldots \quad \forall x \geq 4$$

Where $r = 1,2,3,4,5,6,7,8,9$ y 10. The third term, between brackets [ ] or right parentheses, corresponds to the filling of the orbitals d and the parenthesis (x - 4) indicates that this filling occurs from period 4 or when the subshell 4s has been filled. We make the substitution $x = 4,5,6,7$ in this function and it results for

$x = 4:$ $Z = 20 + r$ $x = 5:$ $Z = 38 + r$
$x = 6:$ $Z = 70 + r$ $x = 7:$ $Z = 102 + r$

But the atomic numbers 20, 38, 70 and 102 are obtained from the function (1.12) for fully filled subshell f, r = 14, that is, the function

$$f_{14}(x) = 2x + 6(x-2) + 10(x-4) + 14(x-5)$$

With which we have in general $r = Z - f_{14}(x)$ to be filling the orbitals d one by one which tentatively indicates a configuration $(x-1)d^r$ $\forall x \geq 4$, Thus, for example, in period x = 4 of the Periodic Table, figure 1.3, we find the elements of atomic numbers 21, 22, 23, 24, 25, 26, 27, 28, 29, and 30 if r = 1, 2, 3, 4, 5, 6, 7, 8, 9 and 10, respectively. Consequently, apart from the rest of the electrons, in the Periodic Table for its atomic number and column we would have for the orbitals d: $3d^1, 3d^2, 3d^3, ..., 3d^{10}$ for Z = 21, 22, 23, ..., 30, respectively. Let's see what follows. The atoms have a shell of noble gas with fully filled p-subshells and at each quantum level n the subshell ns always begins to fill. This creates a large energy difference between each subshell s and the preceding p-subshell. This energy difference has as a consequence that the filled subshell does not produce external field and their charge distributions are said to have spherical symmetry. All this leads us to subtract from the total configuration the noble gas shell and analyze the rest of the electronic distribution. A very ingenious way is through the diagrams that I call **BDED [Barrier Diagram (noble gas) and Electronic Gates].** To illustrate the above, we will determine the electronic configuration of the platinum element whose atomic number is equal to $Z = 78$ (period x = 6 and column r = 8). The function for the noble gas shell (1.16) $\left[p(x-1) \right]$ gives us the noble gas before the period x = 6:

$$\left[p(x-1) \right] = 2(x-1) + 6(x-2) + 10(x-4) + 14(x-6)$$

$$\left[p(6-1) \right] = \left[p(5) \right] = 2(6-1) + 6(6-2) + 10(6-4) = 54 \; (Xe)$$

The Xe closes the cycle in the period $(6-1) = 5$ and is the shell of $Z = 78$. We already have the shield of noble gas we lack the rest of the electrons, that is, 78 - 54 = 24; we subtract two electrons for the orbital $6s^2$ because each period starts in a new subshell **xs**. Now we return to the function (1.10) We substitute x = 6 and r = 8 that it gives us exactly $Z = 78$ and, in turn, it will give us the tentative electronic configuration for platinum, Pt:

$$d(6)_{r=8} = 2(6) + 6(6-2) + \left[10(6-4) + r \right] + 14(6-5)$$

$1s^2$

$2s^2$ $2p^6$

$3s^2$ $3p^6$ $3d^{10}$

$4s^2$ $4p^6$ $4d^{10}$ $\boxed{4f^{14}}$

$5s^2$ $5p^6$ $\boxed{5d^8}$

$\boxed{6s^2}$

The function for the Xe gives us $1s^2 2s^2 3s^2 4s^2 5s^2 2p^6 3p^6 4p^6 5p^6 3d^{10} 4d^{10}$ that in all BDED, and for all noble gas, we lock in continuous line and we have left $4f^{14} 5d^8 6s^2$. All this information still does not tell us how electronic transit should happen. However, one way is by placing this information in a diagram where we can enclose the noble gas shell (BARRIER) and analyze the rest of the electrons in the subshells that can indicate the transit from one sublayer to the other (ELECTRONIC DOORS) as it is presented below:

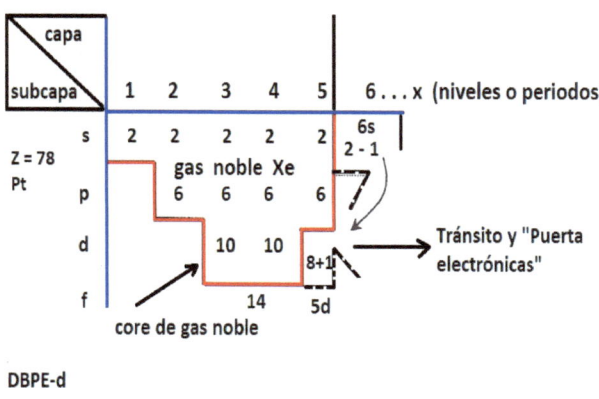

Figure 1.7 The arrow indicates the ***only possibility of electronic transit*** of the 6s orbital, which does not admit electrons due to the exclusion principle, towards the 5d subshell. The analysis of the selection and symmetry [3] rules demonstrate the important conclusion according to which **electronic transitions are only allowed between odd and even states,** as in this case. In addition, from the experimental observations, it is deduced that the energy differences are very small between the subshells 3d and 4s, 4d and 5s, and 5d and 6s and the matter is that the total energy of the atom is the smallest. Such a situation occurs between subshells 5d and 4f, and 6d and 5f. These diagrams allow us to visualize where electronic transits can occur and to deduce mathematical functions for configurations.

According to the BDED-d the electronic doors are between the subshell 6s and 5d and the reason for this is that, firstly, the subshell 4f is completely full; secondly, the spectroscopic calculations for the selection rules lead to the important conclusion according to which **electronic transitions are allowed only between odd and even states** (see the cited bibliography). Here, as the diagram is drawn up, it indicates, from what has already been said, that the transition will occur from sublayer 6s to 5d;. the transition occurs between an **even and odd state**. In addition, the principle of exclusion does not allow more than two electrons per orbital so that the transition goes from 6s to 5d.

With all this information, the function for the electronic configuration of some B-d elements is immediately deduced. We represent this function, only for four blocks, as follows:

$$[Z]_{d_r(x)} = [2(x-1) + 6(x-2) + 10(x-4) + 14(x-6)] \times s^{2-y} (x-2) f^{14}$$

$$(x-1)d \{2x+6(x-2)+[10(x-4)+r]+14(x-5)\} - \{2x+6(x-2)+10(x-4)+14(x-5)\}+y$$

(1.19)

We have written in the order of completion. The final form in the increasing order of levels and simplified is as follows:

$$CE[Z]_{d_r}(x) = [p(x-1)].(x-2)f^{14}.(x-1).d^{[Z-(f_{14}(x))]+y} . x \, s^{2-y}$$

(1.20)

$CE[Z]_{d_r(x)}$ **Read**: electronic configuration of the element of atomic number Z of the Block-d in the column $1 \le r \le 10$ of the period x.

$$Z = 2x + 6(x-2) + [10(x-4) + r] + 14(x-5)$$

In conclusion: for an element located in block d in the column r = 8, period x = 6 with a differential electron y = 1 the electronic function (1.19) gives us the order of filling

$$CE\,[78]_{d_8}(6) = [Xe] \cdot 6s^1 \cdot 4f^{14} \cdot 5d^9$$

Increasingly, levels remain

$$CE\,[78]_{d_8}(6) = [Xe] \cdot 4f^4 \cdot 5d^9 \cdot 6s^1$$

Note:
For multielectronic atoms, the configurations are determined experimentally in the gaseous state, and they are not necessarily retained in the other states. The readings of the spectrum are very complex to have electronic transitions, so it is not always possible to assign the correct configuration to the atom without any ambiguity. Therefore, the electronic configurations are approximate according to the available data and there are cases where it is not clear which is the best configuration [see Sienko / Plane, Theoretical and Descriptive Chemistry, page 449, Editorial Aguilar, 1976].
In addition, deviations for some elements of the d and f blocks should not be given too much attention since they are not very interesting from the chemical point of view [see, Huheey's Inorganic Chemistry, 2nd Edition, Editorial Harla, and Analytical Chemistry, Skog / West, 4th Edition, McGraw Hill, P. 452].

§7.4 Function for the Electronic Configuration of the Elements that fill the orbitals f: Block-f

To obtain the function for the configurations of the elements of the B-f we base ourselves on the following BDED-f:

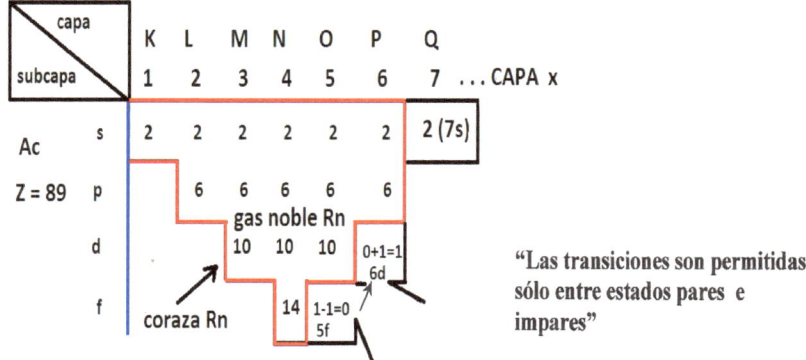

Figure 1.8 Here it is the element Ac (Z=89) of B-f in the period x = 7 and r = 1. According to the experiment, with $y = 1 \text{ ó } 2$ electrons that pass from $(x-2) f^{p-y} \rightarrow (x-1) d^{y}$ **(here y = 1)**. The calculation of the rules of selection and symmetry [3] in quantum mechanics leads to transitions between an odd state (n = 5) to an even state (n = 6) as seen here reflected.

The reasoning made for the function of the electronic configuration of the B-d is analogous to the B-f that we are considering here. The result for the function that gives us the electronic configuration of the elements of block f according to the experiment is as follows:

$$CE[z]_{f_r}(x) = [p(x-1)] \cdot (x-2) f^{[z - g_{18}(x)] - y} \cdot (x-1) d^y \cdot x \, s^2$$

(1.21)

$p = [z - g_{18}(x)]$ are the differential electrons for the orbital f once the s or g orbitals are filled (if they exist or are created), the function $g_{18}(x)$ being (which the reader can deduce) the following:

$$g_{18}(x) = \underbrace{[2x + 6(x-2) + 10(x-4) + 14(x-6)]}_{\rightarrow \text{ part that matches the function of block s } \leftarrow} + 18(x-7)$$

(1.22)

Substituting in (1.21) $x = 7$ $r = 1$ $y = 1$ results the CE:

$[Rn]5f^0 6d^1 7s^2$ for Z=89 and corresponds to the element Ac.

§ 8. Functions for the periodic length (elements in each period)

Each period in the Periodic Table starts with a subshell ns $\forall n \geq 1$ and ends in the np subshell $\forall n \geq 2$. This results in the fact that the atomic number Z, B-s, for all the elements in r = 2 of a shell (level) n a otro $(n-1)$ will carry out the complete cycle in the subshell $(n-1)$s to ns (note that we can go from right to left or vice versa, backwards). The cycle is equal to the periodic length or the number of elements covered by that period. Take, for example, the elements of Z = 2, 4, 12, 20, 38, 56, 88 and 120 in the periods, respectively, x = 1, 2, 3, 4, 5, 6, 7 and 8. Each period starts with an orbital s which implies, for example, If we make differences between a period and the previous one, we have: $120(x=8) - 88(x=7) = 32$ which is the length in the period before the period x = 8, that is x = 7; $88(x=7) - 56(x=6) = 32$ length in x = 6. We find that in the periods x = 6 and 7 have length equal to 32 elements. We continue making the differences until the period x = 0:

$56(x=6) - 38(x=5) = 18$ elements in the period $x=5$

$38(x=5) - 20(x=4) = 18$ elements in the period $x=4$

$20(x=4) - 12(x=3) = 8$ elements in the period $x=3$

$12(x=3) - 4(x=2) = 8$ elements in the period $x=2$

$4(x=2) - 2(x=1) = 2$ elements in the period $x=1$

$2(x=1) - 0(x=0) = 2$ elements in the period $x=0$

IMPORTANT: Article 4: "Algebra and Geometry of Periodic Systems of Elements", published on Amazon, gives the development that leads to a period and level zero and, in addition, the extensive calculation of algebra and geometry of periodic systems. The connection between the quantum mechanics of hydrogen-like atoms and the algebra and periodic geometry, totally unpublished, is found in Articles 2 and 3 in Amazon, and will later be summarized in a single Article.

Note that the height is obtained in the period on the left figure 1.2. We find that height 2, 8, 18, and 32 are repeated in the consecutive even and odd periods: (0,1); (2,3); (4.5) and (6.7), respectively. Now we find that the Z = 2, 4, 12, 20, 38, 56, 88 and 120 are calculated for the respective periods x = 1, 2, 3, 4, 5, 6, 7 and 8 with the function (1.14) that we copy next

$$s_2(x) = \sum_{l=0}^{n-1} 2(2l+1)[x-2l] \quad (1.23)$$

If we call $T(x) =$ longitud periódica $\forall x \geq 0$ (figure 1.3) it is going to fulfill, as previously considered, that $T(x) = s_2(x+1) - s_2(x)$, that is

$$T(x) = \left\{ 2(x+1) + \sum_{L=1}^{n-1} 2(2l+1)[x-(2l-1)] \right\} - \left\{ 2x + \sum_{l=1}^{n-1} 2(2l+1)[x-2l] \right\}$$

(1.24)

We develop for four blocks that corresponds to the maximum for the angular momentum $l = 3$ $(n = 4)$ and the function for the number of elements in the Periodic Table, including x = 0, results:

$$T(x) = \{2(x+1) + 6(x-1) + 10(x-3) + 14(x-5)\} - \{2x + 6(x-2) + 10(x-4) + 14(x-6)\}$$

(1.25)

The function (1.25) allows to obtain: **1)** the number of elements in each period or the periodic length; **2)** the type of subshells that are filling the elements in the corresponding period. Let's see through the following example.

Example 1.4

> With the function (1.25) calculate the length and type of subshells of the elements that form the period x = 4 in the Periodic Table (In figure 1.2 for column and figure 1.3 for row)

We substitute x = 4 (fourth period), and remembering that the parentheses are greater than zero we have:

$$T(x) = \{2(x+1)+6(x-1)+10(x-3)+14(x-5)\} - \{2x+6(x-2)+10(x-4)+14(x-6)\}$$

$$T(4) = \{2(4+1)+6(4-1)+10(4-3)\} - \{2(4)+6(4-2)\} = 38(Sr) - 20(Ca) = 18$$

$$\begin{Bmatrix} 0s^2 \\ 1s^2 \\ 2s^2 & 2p^6 \\ 3s^2 & 3p^6 & 3d^{10} \\ 4s^2 & 4p^6 \end{Bmatrix} - \begin{Bmatrix} 0s^2 \\ 1s^2 \\ 2s^2 & 2p^6 \\ 3s^2 & 3p^6 \end{Bmatrix} = 4s^2 \quad 3d^{10} \quad 4p^6$$

Physically it means that period 4 has a length of 18 and the orbitals used in the construction of these fully filled subshells elements correspond to $4s^2 \; 3d^{10} \; 4p^6$. Note that starts counting the orbital s in the period x = 0 whose height is two to make the difference between Sr (Z = 38) and Ca (Z = 20) so that, as explained, the length in the previous period ax = 5, that is, x = 4 (see the differences above). This is general for any period. In later articles we will give specific details.

Let's take a closer look at the relationship from period zero which must go in the Periodic Table, and in general in any periodic system. We start from the relationship (1.24) only for orbitals and multiply by two according to the principle of exclusion: we have to $T(x) = s_2(x+1) - s_2(x)$ where

$$s(x+1) = \{(x+1) + 3(x-1) + 5(x-3) + 7(x-5)\} \text{ and}$$

$$s(x) = \{x + 3(x-2) + 5(x-4) + 7(x-6)\}$$

Period zero: x = 0 (l = 0, n=1)

$$s(0+1) = s(1) = \{(0+1)\} = 0s^1 \Rightarrow 2(0s^1) = 0s^2$$

in the periodic table, Figure 1.3, corresponds to two squares: -1 and 0. The interpretation of the square -1 can be in the polarization of the vacuum (it is outside of quantum algebra and remains for experts in the area). Regarding the zero square it is clear that it is the interaction of the wave **s** (**l=0**) with the nucleus. Now for s(x), x = 0, result s(0) = 0 (box 0 code, Z=0). At last:

$$T(0) = \{2(0+1)\} - \{2 \times 0\} = 2$$

$$\left\{ 2 \begin{vmatrix} 0 \\ 0s^1 \end{vmatrix} \right\} - \{0\} = 0s^2$$

Where $0s^2$ represents the two labeled squares -1, 0.

Period one: x = 1 (l = 0, n=1)

For orbital and then multiply by two:

$$s(1+1) = s(2) = \{(1+1)\} = 0s^1 + 1s^1 \Rightarrow 2(0s^1 + 1s^1) = 4 \text{ squares from } x = 0$$

Counted from period zero. Now $s(1) = 1 = 0s^1$ **(calculated in period zero)**. At last:

$$T(1) = \{2(1+1)\} - \{2 \cdot 1\} = 2$$

$$\left\{ 2 \begin{vmatrix} 0s^1 \\ 1s^1 \end{vmatrix} \right\} - 2\{0s^1\} = 1s^2 \quad \text{two elements in X = 1: H, He}$$

So on.

42

§ 8.1 Periodic length function obtained from the relationship that determines the energy levels of hydrogen atoms that leads to the Periodic Table.

The experimental results, mainly the analysis of the spectra of the atoms with an equal number of electrons that of protons, or Z electrons, reveal similarities with the spectrum of the hydrogen atom. In the introduction to this paper, we present some similarities in paragraphs A, B, C and D. We can add that:

(1) the elements of the Periodic Table, PT, have stationary states only for certain energy values.

(2) The spectra of these elements are explained based on transitions between levels as in the atom of H.

(3) The diagram of energy levels of H is similar to that of polyelectronic atoms, these atoms still have n^2 levels for each value of n but they no longer have the same energy or are no longer degenerate.

<u>**The length in the periods of the PSCE**</u> can be obtained from the mechano-quantum study of hydrogen atoms as explained below.

Relationship that determines the distribution of energy levels that leads to the PSCE (Periodic Table)

In the study of motion in a central [4] symmetric field, in the case of atoms with Z protons and a single electron, the Schroedinger equation for the stationary states transformed into spherical coordinates admits to be separated into a radial and an angular part. The states that belong to the energy spectrum are obtained from the analysis of the radial part and it is found that the relationship that determines the distribution of the energy levels of the hydrogen atom vienna given by

$$n_r + l + 1 = \frac{Z}{ac}, \qquad a = \frac{\hbar^2}{me^2} \qquad \text{and} \qquad c = \sqrt{\frac{-2E_n}{ae^2}} \qquad (1.26)$$

Here $\hbar, m$ y e are the known constants of atomic physics. . This relationship is expressed differently according to the conditions given by the quantum numbers n_r, l y n: $n_r + l + 1 = n$ ∧ $n = l + 1$ consequently (1.26) it is as follows

$$Za^{-1} - c - cl = 0 \quad [1] \qquad (1.27)$$

And the relation (1.27), equivalent to (1.26), **determines the distribution of the energy levels of atoms with Z protons and one electron.**

Now let's see how this relationship leads us to determine the length in the periods of the psce or periodic table

To do this, let's look at figure 1.2. Note that the height (or length in figure 1.3) is repeated in an even period, x_0, and odd, x_f, for the maximum value of the quantum number of angular momentum $l = 0, 1, 2, 3, \ldots n - 1$. For example, if we go to the level n = 3, the periods $x_0 = 4$ y $x_f = 5$ are filling the maximum number of orbitals $l = 2$. But we found that there is a relationship for the Periodic Table between the quantum number of maximum angular momentum and the periods x_0 y x_f that is met for the levels n = 1, 2, 3 and 4 in figure 1.2 [actually there is a general relationship for any system newspaper since the PSCE is not unique]. This relationship is given as follows:

$$x_0 = 2l \quad y \quad x_f = 2l+1 \quad \forall l = 0,1,2,3,\ldots,n-1 \quad (1.28)$$

For even periods $x_0 = 0,2,4,6,\ldots$ and the odd $x_f = 1,3,5,7,\ldots$, it is as shown in figure 1.2. What is done now? We clear the quantum number ℓ of (1.28) and substitute in (1.27). Let's apply for even periods (for odd periods it's similar): $l = x_0/2$, it remains:

$$Za^{-1} - c - c\left(\frac{x_0}{2}\right) = 0 \quad \Rightarrow \quad \frac{Z}{ac} = \frac{(x_0+2)}{2}$$

By replacing the constants given in (1.26) and clearing the energy, it finally results in x_0 y x_f (see Example 1.4 and details below):

$$E_{x_0} = -\frac{13.6}{\left[\frac{x_0+2}{4}\right]^2}\,eV \; si \; x_0 = 2l \quad \wedge \quad E_{x_f} = -\frac{13.6}{\left[\frac{x_f+1}{4}\right]^2}\,eV \; si \; x_f = 2l+1$$

$$(1.29)$$

In both cases when replacing (1.28) it is reduced to the expression $E_n = -Z^2\dfrac{13.6\,eV}{n^2}$, which is the Bohr relation for hydrogen (see note on calculating energy levels for hydrogen).

Of the relation (1.29) **the denominator** is of interest. This gives us the number of orbitals in each even and odd period; we now assign **two**

electrons per orbital (exclusion principle) and the denominator is a function $T(x_0)$, $T(x)$ for the height in the even and odd periods, respectively, in figure 1.2 including the period x = 0 and it results: $(x \equiv x_f)$

$$T(x_0) = \frac{(x_0+2)^2}{2}, \quad \forall x_0 = 0, 2, 4, 6, \ldots, 2l \qquad T(x_f) = \frac{(x_f+1)^2}{2}, \quad \forall x_f = 1, 3, 5, 7, \ldots, 2l+1$$

(1.30)

instead, the function (1.25) determines the height in the PSCE, figure1.2, including all the periods. The function (1.30) calculates separately the height for the odd and even periods. Now consider the following: in (1.30) for the electronic wave s (electronic angular momentum l = 0) leads us to the levels of the atom 0 and 1 (see figures 1.1-1.4) and to match the periods $x_0 = 0$ y $x = 1$ the axis of the parabola is the one that corresponds to the function $T(x) = (x+1)^2/2$, that is, $x = -1$ so that the integration is made from the vacuum of interaction of the electronic wave s with the nucleus what actually happens and leads to physical reality

otherwise we would obtain absurd results if we did not start from scratch. Furthermore, in the calculation of the periodic functions for each block the integration is always **carried out taking the previous level** to a new level because the electronic wave travels through the vacuum between one level and another. It is clear that from the relation (1.27) the filling sequence is constructed of the periodic table where the hydrogen atom is the raw material from which the chemical elements are built (The universe is made up of 92% hydrogen nuclei and about 8% helium and the remaining elements of the Periodic Table just 0.1%, according to astrophysicists). Another aspect of interest is that in the denominators of

(1.29) there is no problem in assigning more than two electrons to the atomic orbitals and the resulting system still Has a mathematical solution only the design parameter is no longer ½, which is the one corresponding to the Periodic Table, but one according to how the filling sequence is chosen.

Note on the calculation of energy levels

In the relation (4.29) appears the constants a and c which are:

$$a = \frac{\hbar^2}{\mu e^2} \quad (a = 0.529 \text{ Å}) \quad y \quad c \equiv \left(\frac{-2E}{ae^2}\right)^{\frac{1}{2}}$$

Where

$e = 4.80325 \times 10^{10} cm^{3/2} g^{1/2} seg^{-1}$ (unidades electrostáticas)

$\hbar = 1.05488 \times 10^{-27} g \times cm^2 \times seg^{-1}$ $(ergio \times seg) \equiv (g \frac{cm^2}{seg^2} \times seg)$

Masa reducida, $\mu = 9.104 \times 10^{-28} g$

When replaced in (1.27) l=n-1 we clear E, you have:

$$E_n = -\frac{Z^2 \mu e^4}{2\hbar^2} \frac{1}{n^2} = -\frac{4\,845.\,88401 \times 10^{-68}}{2.\,225544 \times 10^{-54}} \frac{g^3 cm^6 seg^{-4}}{g^2 cm^4 seg^{-2}} \frac{1}{n^2} = -\frac{2.178 \times 10^{-11} Z^2}{n^2} ergios$$

We move on to (eV): $(1 eV = 1,6022 \times 10^{-19} julios)(1 julio = 10^7 ergios)$

$$E_n = -\frac{Z^2 13.\,5941}{n^2} eV \quad = \quad -\frac{Z^2 13.\,6}{n^2} eV$$

Relevant information: Here and in all articles the dot is decimal. For units of a thousand we will use comma as in the Anglo-Saxon nomenclature or we will separate the digits by placing the period for the decimal part.

Next, we will go further to find a single relation for (1.30) and its application to the Periodic Table. The starting point is the function for the orbitals corresponding to the s-states (1.14) that we rewrite:

$$s(x) = \sum_{l=0}^{n-1} (2l+1)[x - 2l] \quad (1.31)$$

In addition, we have to

$$s(x+1) = \sum_{l=0}^{n-1} (2l+1)[x - (2l-1)] \quad (1.32)$$

As explained, the difference $T(x) = s(x+1) - s(x)$ represents the number of squares (orbital functions) in periods $x = 0, 1, 2, \ldots$ to be filled with two electrons (excluding $x = 0$) to design the Periodic Table (see Figure 1.9), i.e.

$$T(x) = \left\{ \sum_{l=0}^{n-1} (2l+1)[x - (2l-1)] \right\} - \left\{ \sum_{l=0}^{n-1} (2l+1)[x - 2l] \right\}$$
(1.33)

This function gives for periods $\alpha \equiv x = 0, 1, 2, 3, 4, 5, 6, 7, 8, \ldots$ the number of orbitals (squares), respectively, 1, 1, 4, 4, 9, 9, 16, 16, 25,… for the filling that will lead to the Periodic Table by assigning two electrons per orbital (exclude period zero that has already been explained. Electrons are placed from period one.)

Let's develop (1.33):

$$T(x) = \left\{ x + \sum_{l=0}^{n-1}(2l+1)[x-2l] + \sum_{l=1}^{n-1}(2l+1)+1 \right\} - \left\{ x + \sum_{l=1}^{n-1}(2l+1)[x-2l] \right\}$$

canceling equal term, remains

$$T(x) = \sum_{l=1}^{n-1}(2l+1)+1 = \sum_{l=0}^{n-1}(2l+1) = n^2 = (l+1)^2, \forall l = 0,1,2,\ldots,(n-1)$$

It follows from this that

$$T(x) = (l+1)^2, \forall l = 0,1,2,\ldots,(n-1) \Rightarrow l = \sqrt{T(x)} - 1 \quad (1.34)$$

We substitute (1.34) in (1.27):

Note: Values of the physical constants appear in the box on page 47.

$$Za^{-1} - c - c\left[\sqrt{T(x)} - 1\right] = Za^{-1} - \not{c} - c\sqrt{T(x)} + \not{c} = 0 \Rightarrow Za^{-1} = c\sqrt{T(x)} \Rightarrow$$

$$\frac{Z}{ac} = \sqrt{T(x)} \quad \Rightarrow \quad E_n = -\frac{Z^2 \times 13.6}{T(x)} eV$$

Now we substitute (1.33), it is finally the orbital energy of the electrons to design the Periodic Table:

$$E_x = -\cfrac{Z^2 \times 13.6 \text{ eV}}{\left\{(x+1) + \sum_{l=1}^{n-1}(2l+1)\left[x - (2l-1)\right]\right\} - \left\{\sum_{l=0}^{n-1}(2l+1)\left[x - 2l\right]\right\}}$$

(1.35)

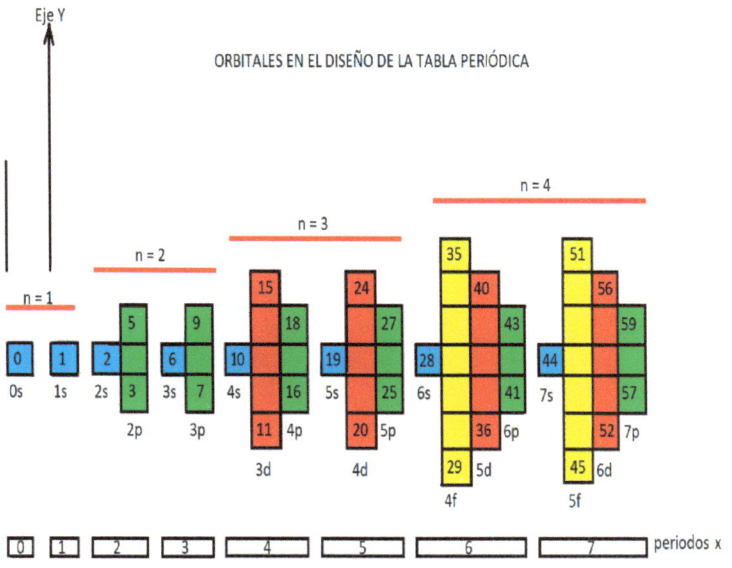

Función orbital = p(x) = x + 3(x-1) + 5(x-3) + 7(x-5), x corresponde con el periodo: x = 0, 1, 2, 3, 4, 5, 6 y 7   (hasta n = 4)

Figure 1.9 This is the sequence that leads to the Periodic Table of Elements by placing two electrons in each square from 1s (see Figure 1.1).

We develop (1.35) up to $n=4$ ($l=3$) [the n of the staggered function. Notice how the periodic variable x overlaps with level n] we have left:

$$E_x = -\cfrac{Z^2 \times 13.6 \text{ eV}}{\left\{(x+1) + 3(x-1) + 5(x-3) + 7(x-5)\right\} - \left\{x + 3(x-2) + 5(x-4) + 7(x-6)\right\}}$$

(1.36)

Example 1.5

Calculate the orbital energy of the electron placed in the period x = 5 under the level n = 3 of (Figure 1.9) and what orbitals it can occupy. What is its physical significance?

Solution

This is an odd period: $5 = 2l + 1 \Rightarrow l = 2 \, (n = 3)$, is the n=3 of the step function in Figure 1.9. there we see that the orbitals where the electron can be placed are the 5s4d5p. The energy is calculated with the ratio (1.29):

$$E_{x=5} = -\frac{13.6}{\frac{[5+1]^2}{4}} eV = -\frac{Z^2 \times 13.6}{3^2}$$

Note that this is the Bohr relation for n=3 where the denominator gives us 9 orbitals that would be $3s^13p^33d^5$, that is, an electron 3s3p3d can be placed in an orbital 5s4d5p as the relationship indicates (1.36):

$$E_5 = -\frac{Z^2 \times 13.6 \; eV}{\{(5+1)+3(5-1)+5(5-3)\}-\{5+3(5-2)+5(5-4)\}} = -\frac{13.6 \; eV}{\{28\}-\{19\}} = -\frac{13.6}{3^2} eV$$

We get the same result, but there is a difference and here more information, which one? Now let's just go to the denominator of this energy relationship.

$$T(x) = \{(x+1)+3(x-1)+5(x-3)\}-\{x+3(x-2)+5(x-4)\}$$
$$T(5) = \{(5+1)+3(4)+5(2)\}-\{5+3(3)+5(1)\} = \{6+12+10\}-\{5+9+5\} = 9$$

$$\begin{Bmatrix} 0s^1 \\ 1s^1 \\ 2s^1 \quad 2p^3 \\ 3s^1 \quad 3p^3 \quad 3d^5 \\ 4s^1 \quad 4p^3 \quad 4d^5 \\ 5s^1 \quad 5p^3 \end{Bmatrix} - \begin{Bmatrix} 0s^1 \\ 1s^1 \\ 2s^1 \quad 2p^3 \\ 3s^1 \quad 3p^3 \quad 3d^5 \\ 4s^1 \quad 4p^3 \end{Bmatrix} = 5s^1 4d^5 5p^3 = 9 \text{ orbitales}$$

This result is observed in Figure 1.9 where we count the unit area grids from the zero period, whose grid is labeled zero, up to x = 5 there are 28 grids: 6 states s + 12 states p + 10 states d = 28 squares. We now count the squares from x = 0 to x = 4: 5 states s + 9 states p + 5 states d = 19 squares (u orbital states). Finally, we subtract, it turns out $5s^1 4d^5 5p^3$.

And we found additional information that does not supply the relationship of quantum mechanics. This result tells us that an electron 3s, 3p and/or (y/o) 3d of energy $-13.6/9$ eV can be placed in any of the orbitals of the sublayers 5s, 4d y/o 5p despite being carried to orbitals of levels above level n = 3 (the n of the staggered function).

How to apply them to the construction of the elements of the Periodic Table? Here it is something interesting. We must substitute in the orbital function of energy (1.36) the function corresponding to each block of elements as we will see in the next paragraph.

§ 8.2 Orbital energies and filling functions for s-block elements

The following function refers to a single electron for a nucleus of atomic number Z that corresponds to the elements of the s-block that initiate each period of the Periodic Table. The numerator refers to both the atomic number of the element and its electronic distribution. The denominator calculates the length or number of elements that will fill each period, as well as the type of orbital; the length starts precisely in the block of elements that are filling the orbitals **s**. The length of the block **s** is $r = 2$, that is, this block is formed by two rows or two columns according to the vertical or horizontal table as it appears in the figures 1.2 y 1.3, respectively.

$$E_{x,r} = -\frac{\{2(x-1)+r\}+6(x-2)+10(x-4)+14(x-6)\}^2 \times 13.6 \text{ eV}}{\{(x+1)+3(x-1)+5(x-3)+7(x-5)\}\cdot\{x+3(x-2)+5(x-4)+7(x-6)\}}$$

(1.37)

How do we apply this relationship to the Periodic Table given in Figure 1.2 y/o 1.3?

We are applying period x = 0 for r = 1.2; then period x = 1, r = 1.2; so on. That is, period x, column r of the block corresponding to the filling of the orbitals s, p, d and f of the Periodic Table.

Level n = 1 periods x = 0, 1 and r = 1.2

Period: $x = 0, \ r = 1$ $(l = 0, n = 1)$ **Z = -1**

$$E_{0,1} = -\frac{\left\{2(0-1)+1\right\}^2 \times 13.6 \text{ eV}}{\left\{0+1\right\}-\left\{0\right\}} = -\frac{\left\{-1\right\}^2 \times 13.6 \text{ eV}}{1(n=1)} = -13.6 \text{ eV } (Z=-1)$$

$$\left|\begin{array}{c}0\\0s^1\end{array}\right| - |0| = 0s^1$$

Period: $x = 0$, $r = 2$ $(l = 0, n = 1)$ $Z = 0$

$$E_{0,2} = -\frac{\left\{2(0-1)+2\right\}^2 \times 13.6 \text{ eV}}{\left\{0+1\right\}-\left\{0\right\}} = -\frac{\left\{0\right\}^2 \times 13.6 \text{ eV}}{1(n=1)} = 0 \text{ } (Z=0)$$

$$\left|\begin{array}{c}0\\0s^1\end{array}\right| - |0| = 0s^1$$

The calculation leads to the conclusion of including a zero period in the Periodic Table. Article 4 makes the most elaborate and extensive calculation from the raw material hydrogen in the formation of the chemical elements. To construct the Periodic Table, all the properties of the periodic system of hydrogen atoms are duplicated despite the calculation made for a nucleus of Z protons and a single electron. Note in the relation (1.37) the denominator refers to the atomic numbers and their configuration of the elements of the block of orbitals s that are at the end of the text.

Period: $x = 1$, $r = 1$ $(l = 0, n = 1)$ $Z = 1$ (H)

$$E_{1,1} = -\frac{\left\{2(1-1)+1\right\}^2 \times 13.6 \text{ eV}}{\left\{1+1\right\}-\left\{1\right\}} = -\frac{\left\{+1\right\}^2 \times 13.6 \text{ eV}}{1(n=1)} = -13.6 \text{ eV } (Z=+1, \text{ H})$$

$$\left|\begin{array}{c}0s^1\\1s^1\end{array}\right| - |0s^1| = 1s^1 \text{ (orbital 1s inicia el llenado con 2 electrones)}$$

Period: $x = 1$, $r = 2$ $(l = 0, n = 1)$ $Z = 2$ (He)

$$E_{1,2} = -\frac{\{2(1-1)+2]\}^2 \times 13.6 \text{ eV}}{\{(1+1)\}\{1\}} = -\frac{\{-1\}^2 \times 13.6 \text{ eV}}{1(n=1)} = -13.6 \text{ eV} \left(Z = 2, He\right)$$

$$\begin{vmatrix} 0s^1 \\ 1s^1 \end{vmatrix} - |0s^1| = 1s^1 \Rightarrow 2(1s^1) = 1s^2 \; (He)$$

Conclusion

The calculation corresponds under the level n = 1 (Figure 1.2) comprising periods 0 and 1 (x = 0 and x = 1). Note that in period zero a labeled "something" appears Z =-1, What to attribute this particle to? This escapes the ends of this work; however, one explanation may lie in the nature of the vacuum and **the polarization of the vacuum:** see references 7 and 8. In addition, the calculation gives for x = 0 two squares that have been explained above. For the period x = 1, which is where the filling of the orbital 1s with electrons begins, we have the elements of period one of the Table that are H and He of configuration found in the numerator 1s¹ y 1s², respectively. Note that the maximum value of the angular impulse corresponds to l = 0 for n = 1.

Ahora pasamos a analizar los elementos del bloque s bajo el nivel n = 2 de la función escalonada en los periodos 2 y 3.

Level n = 2 periods x = 2.3 and r = 1.2

Period: $x = 2$, $r = 1$ $(l = 1, n = 2)$ $Z = 3$ (Li)

$$E_{2,1} = -\frac{\{2(2-1)+1]\}^2 \times 13.6 \text{ eV}}{\{(2+1)+3(1)\}-\{2(1)\}} = -\frac{\{3\}^2 \times 13.6 \text{ eV}}{4(n=2)}, \quad Z = 3\,(Li)$$

$$\begin{vmatrix} 0s^1 \\ 1s^1 \\ 2s^1 \quad 2p^3 \end{vmatrix} - \begin{vmatrix} 0s^1 \\ 1s^1 \end{vmatrix} = 2s^1 \quad 2p^3 \Rightarrow \text{longitud } T(2) = 1 : 2s^1$$

The numerator gives us the configuration 1s² 2s¹ which corresponds to the element Li of atomic number Z = 3. The denominator corresponds to the periodic length function, T(x), ratio (1.25) when assigning two electrons per orbital, i.e. 2(2s1 2p3) = 8 elements in period x = 2 when the 2s and 2p orbitals have been filled; however, we are starting the period x = 2 by using only the 2s¹ orbital which gives us a length T(2)=1.

Period: $x = 2$, $r = 2$ $(l = 1, n = 2)$ $Z = 4$ (Be)

$$E_{2,2} = -\frac{\{2(2-1)+2\}^2 \times 13.6 \text{ eV}}{\{2+1\}+3(1)\}\{2(1)\}} = -\frac{\{1s^2 2s^2\}^2 \times 13.6 \text{ eV}}{2^2 (n=2)}, \quad Z = 4 \text{ (Be)}$$

$$\begin{vmatrix} 0s^1 \\ 1s^1 \\ 2s^1 \quad 2p^3 \end{vmatrix} - \begin{vmatrix} 0s^1 \\ 1s^1 \end{vmatrix} = 2s^1 \quad 2p^3 \Rightarrow \text{longitud } T(2) = 1 : 2s^2$$

Note that the energy to place an electron in the nuclei Z = 3.4 coincides the period x = 2 with the level n = 2. However, below, we will see that this does not happen in the following periods.

Period: $x = 3$, $r = 1$ $(l = 1, n = 2)$ $Z = 11$ (Na)

$$E_{3,1} = -\frac{\{2(2)+1]+6(1)\}^2 \times 13.6 \text{ eV}}{\{3+1\}+3(2)\}\{3+3(1)\}} = -\frac{\{1s^2 2s^2 2p^6 3s^1\}^2 \times 13.6 \text{ eV}}{2^2 (n=2)}, \quad Z = 11$$

$$\begin{vmatrix} 0s^1 \\ 1s^1 \\ 2s^1 \quad 2p^3 \\ 3s^1 \quad 3p^3 \end{vmatrix} - \begin{vmatrix} 0s^1 \\ 1s^1 \\ 2s^1 \quad 2p^3 \end{vmatrix} = 3s^1 \quad 3p^3 \Rightarrow \text{longitud } T(3) = 1 : 3s^1$$

Note that the electron is placed in the 3s orbital but with energy n = 2.

Periodo: $x = 3$, $r = 2$ $(l = 1, n = 2)$ $Z = 12$ (Mg)

$$E_{3,2} = -\frac{\{2(2)+2]+6(1)\}^2 \times 13.6 \text{ eV}}{\{3+1)+3(2)\}\{3+3(1)\}} = -\frac{\{1s^2 2s^2 2p^6 3s^2\}^2 \times 13.6 \text{ eV}}{2^2 (n=2)}, \quad Z = 12$$

$$\begin{vmatrix} 0s^1 \\ 1s^1 \\ 2s^1 & 2p^3 \\ 3s^1 & 3p^3 \end{vmatrix} - \begin{vmatrix} 0s^1 \\ 1s^1 \\ 2s^1 & 2p^3 \end{vmatrix} = 3s^1 \quad 3p^3 \Rightarrow \text{longitud } T(3) = 2 : 3s^2$$

The calculation shows that the electron 3s² for a nucleus Z = 12 (Mg) has energy of one electron of H degenerated 2s,2p multiplied by Z; also, note that the length is now 2 elements, Na and Mg. In paragraph 9 we will calculate the path one by one of the elements in the Periodic Table. Note, too, that the filling of the levels in each period 2 and 3 follows the pattern of the corresponding electronic configurations of the elements as they add up from the 1s orbital.

Level n = 3 periods x = 4,5   y   r = 1,2

Period: $x = 4$, $r = 1$ $(l = 2, n = 3)$ $Z = 19$ (K)

$$E_{4,1} = -\frac{\{2(3)+1]+6(2)\}^2 \times 13.6 \text{ eV}}{\{4+1)+3(3)+5(1)\}\{4+3(2)\}} = -\frac{\{1s^2 2s^2 2p^6 3s^2 3p^6 4s^1\}^2 \times 13.6 \text{ eV}}{3^2 (n=3)}$$

$$\begin{vmatrix} 0s^1 \\ 1s^1 \\ 2s^1 & 2p^3 \\ 3s^1 & 3p^3 & 3d^5 \\ 4s^1 & 4p^3 \end{vmatrix} - \begin{vmatrix} 0s^1 \\ 1s^1 \\ 2s^1 & 2p^3 \\ 3s^1 & 3p^3 \end{vmatrix} = 4s^1 \quad 3d^5 \quad 4p^3 \Rightarrow \text{longitud } T(4) = 1 \Rightarrow 4s^1$$

Period: $x = 4$, $r = 2$ ($l = 2, n = 3$) $Z = 20$ (Ca)

$$E_{4,2} = -\frac{\{2(3)+2]+6(2)\}^2 \times 13.6 \text{ eV}}{\{4+1)+3(3)+5(1)\}\{4+3(2)\}} = -\frac{\{1s^2 2s^2 2p^6 3s^2 3p^6 4s^2\}^2 \times 13.6 \text{ eV}}{3^2 (n=3)}$$

$$\begin{vmatrix} 0s^1 \\ 1s^1 \\ 2s^1 & 2p^3 \\ 3s^1 & 3p^3 & 3d^5 \\ 4s^1 & 4p^3 \end{vmatrix} - \begin{vmatrix} 0s^1 \\ 1s^1 \\ 2s^1 & 2p^3 \\ 3s^1 & 3p^3 \end{vmatrix} = 4s^1 \, 3d^5 \, 4p^3 \Rightarrow T(4) = 2(4s^1) \Rightarrow 4s^2$$

Notice that the electron is to be placed in the 4s orbital but with energy n = 3 as indicated by the denominator. Also, note the electronic configuration 1s² 2s² 2p⁶ 3s² 3p⁶ 4s¹,² in the numerator and filling in each given period with the denominator that coincide with the respective distribution of electrons. Let's move on to the next period.

Period: $x = 5$, $r = 1$ ($l = 2, n = 3$) $Z = 37$ (Rb)

Let's be more specific when calculating the atomic number and the electron configuration using the numerator in (1.37) but in the order or sequence of filling because after filling the orbital 4s is filled the orbital 3d and the filling of the orbital 4p and the period 4 is completed to start the filling of the orbital 5s:

$$[2(x-1)+r] + 10(x-4) + 6(x-2)$$

$$[2(5-1)+1] + 10(5-4) + 6(5-2) = 37 \; Rb)$$
$$[2(4)+1] + 10(1) + 6(3)$$

$1s^2$
$2s^2$ $\qquad\qquad\qquad\qquad$ $2p^6$
$3s^2$ $\qquad\qquad\qquad\qquad$ $3p^6$
$4s^2$ $\qquad\quad 3d^{10}$ $\qquad\quad$ $4p^6$
$5s^1$

Note the calculation of filling orbitals s in the denominator and we add from the period x = 1 to the period x = 4 which is exactly this order when the filling of period 4 is completed and begins to fill the orbitals 5s, that is:

$\qquad\qquad 1s^2$ $\qquad\qquad\qquad\qquad\qquad$ periodo: $x = 1$
$\qquad\qquad 2s^2$ $\qquad 2p^6$ $\qquad\qquad\qquad\qquad$ $x = 2$
$\qquad\qquad 3s^2$ $\qquad 3p^6$ $\qquad\qquad\qquad\qquad$ $x = 3$
$\qquad\qquad 4s^2$ $\qquad 3d^{10}$ $\quad 4p^6$ $\qquad\qquad$ $x = 4$
$\qquad\qquad 5s^1$ $\qquad\qquad\qquad\qquad\qquad\qquad$ $x = 5$

Now we calculate the energy of the electron that will occupy the 5s orbital:

$$E_{5,1} = -\frac{\{2(4)+1] + 6(3) + 10(1)\}^2 \times 13.6 \text{ eV}}{\{5+1)+3(4)+5(2)\}\{5+3(3)+5(1)\}} = -\frac{37^2 \times 13.6 \text{ eV}}{3^2 (n=3)}$$

$$\begin{vmatrix} 0s^1 \\ 1s^1 \\ 2s^1 \quad 2p^3 \\ 3s^1 \quad 3p^3 \quad 3d^5 \\ 4s^1 \quad 4p^3 \quad 4d^5 \\ 5s^1 \quad 5p^3 \end{vmatrix} - \begin{vmatrix} 0s^1 \\ 1s^1 \\ 2s^1 \quad 2p^3 \\ 3s^1 \quad 3p^3 \quad 3d^5 \\ 4s^1 \quad 4p^3 \end{vmatrix} = 5s^1 \; 4d^5 \; 5p^3 \Rightarrow T(5) = 1$$

And the quantum-periodic calculation indicates that an electron of H 3s of energy n = 3 is going to be placed in a 5s orbital, how or what or who? But it's the right calculation! The filling of the 5s orbital is completed by adding a differential electron and increasing the nuclear charge by one as specified below.

Period: $x = 5$, $r = 2$ ($l = 2, n = 3$) $Z = 38$ (Sr)

$$\begin{array}{l} \left[2(5-1)+2\right] + 10(5-4) + 6(5-2) = 38 \ Sr) \\ \left[2(4)+1\right] \quad\quad + 10(1) \quad\quad + 6(3) \end{array}$$

$1s^2$
$2s^2$ $\quad\quad\quad\quad\quad\quad\quad\quad\quad$ $2p^6$
$3s^2$ $\quad\quad\quad\quad\quad\quad\quad\quad\quad$ $3p^6$
$4s^2$ $\quad\quad\quad\quad 3d^{10}$ $\quad\quad\quad$ $4p^6$
$5s^2$

Note the calculation of filling the orbitals s in the denominator and we add from the period x = 1 to the period x = 4 which is exactly this order when the filling of period 4 is completed and finish filling the orbitals 5s, that is:

$\quad\quad\quad 1s^2$ $\quad\quad\quad\quad\quad\quad\quad\quad\quad\quad$ periodo: $x = 1$
$\quad\quad\quad 2s^2$ $\quad\quad\quad\quad\quad\quad 2p^6$ $\quad\quad\quad\quad$ $x = 2$
$\quad\quad\quad 3s^2$ $\quad\quad\quad\quad\quad\quad 3p^6$ $\quad\quad\quad\quad$ $x = 3$
$\quad\quad\quad 4s^2$ $\quad\quad 3d^{10}$ $\quad\quad 4p^6$ $\quad\quad\quad\quad$ $x = 4$
$\quad\quad\quad 5s^2$ $\quad\quad\quad\quad\quad\quad\quad\quad\quad\quad\quad$ $x = 5$

Now we calculate the energy of the electron that will occupy the 5s orbital:

$$E_{5,2} = -\frac{\{2(4)+2]+6(3)+10(1)\}^2 \times 13.6 \text{ eV}}{\{5+1)+3(4)+5(2)\}\{5+3(3)+5(1)\}} = -\frac{38^2 \times 13.6 \text{ eV}}{3^2 (n=3)}$$

$$\begin{vmatrix} 0s^1 \\ 1s^1 \\ 2s^1 & 2p^3 \\ 3s^1 & 3p^3 & 3d^5 \\ 4s^1 & 4p^3 & 4d^5 \\ 5s^1 & 5p^3 \end{vmatrix} - \begin{vmatrix} 0s^1 \\ 1s^1 \\ 2s^1 & 2p^3 \\ 3s^1 & 3p^3 & 3d^5 \\ 4s^1 & 4p^3 \end{vmatrix} = 5s^1 4d^5 5p^3 \Rightarrow T(5) = 2(5s^1) = 2$$

And it has been completed to fill the 5s orbital for a length of 2 elements in period 5: T(5) = 2. Again, the quantum-periodic calculation in the design of the Periodic Table indicates that an electron of H of energy 13.6/9 eV (n = 3) can be placed in the orbital s of level n = 5, that is, an orbital 5s.

Note: In Articles 2 and 3, published on amazon, the theory of how this happens is given

Periods 6 and 7 under the staggered function in Figure 1.2 are calculated similarly. Now let's move on to the calculation of the filling of the elements of the orbital block p.

§ 8.3 Orbital energies and filling functions for p-block elements

Next, the form that the energy ratio (1.35) acquires and replace Z with the periodic filling function for the elements of the p orbital block (1.8):

$$E_{x,r} = -\frac{\{2x+[6(x-2)+r]+10(x-3)+14(x-5)\}^2 \times 13.6 \text{ eV}}{\{x+1)+3(x-1)+5(x-3)+7(x-5)\}\{x+3(x-2)+5(x-4)+7(x-6)\}}$$

(1.38)

The filling of the p orbitals starts from the energy level n = 2 of period 2 of the Periodic Table in Figure 1.2.

Level n = 2 periods x = 2,3   y   r = 1,2,3,4,5 y 6

Period: $x = 2$, $r = 1$ ($l = 1, n = 2$) Z = 5 (B)

From the ratio (1.38) we calculate the electronic configuration of the Be:

$$2(2) + [6(2-2)+1]$$
$$1s^2$$
$$2s^2 \qquad 2p^1$$

Note that in period two the 2s orbital is filled and the filling of the 2p orbital for a length begins T(2) = 3. Now we proceed to the calculation of the orbitals to be filled in period two and energy with (1.38):

$$E_{2,1} = -\frac{\{s^2 2s^2 2p^1\}^2 \times 13.6 \text{ eV}}{\{2+1\}+3(1)\}\cdot\{2(1)\}} = -\frac{\{5\}^2 \times 13.6 \text{ eV}}{2^2 (n=2)}, \quad Z = 5(B)$$

$$\begin{vmatrix} 0s^1 \\ 1s^1 \\ 2s^1 \quad 2p^3 \end{vmatrix} - \begin{vmatrix} 0s^1 \\ 1s^1 \end{vmatrix} = 2s^1 \; 2p^3 \Rightarrow \text{longitud } T(2) = 2s^2 2p^1 = 3$$

The process continues until completing 6 electrons which is the capacity of the sublayers p: ($x_0 = 2l$, initial periods and $x_f = 2l + 1$, final periods)

Period: $x = 2$, $r = 6$ ($l = 1, n = 2$) Z = 10 (Ne)

$$E_{2,6} = -\frac{\{s^2 2s^2 2p^6\}^2 \times 13.6 \text{ eV}}{\{2+1)+3(1)\}\{2\}} = -\frac{\{10\}^2 \times 13.6 \text{ eV}}{2^2 (n=2)}, \quad Z = 10 \text{ (Ne)}$$

$$\begin{vmatrix} 0s^1 \\ 1s^1 \\ 2s^1 \quad 2p^3 \end{vmatrix} - \begin{vmatrix} 0s^1 \\ 1s^1 \end{vmatrix} = 2s^1 \ 2p^3 \Rightarrow \text{longitud T}(2) = 2(2s^1 2p^3) = 8$$

And period 2 is completed with 8 elements and the noble gas Ne (Z=10).

Period: $x = 3$, $r = 1$ ($l = 1, n = 2$) $Z = 13$ (Al)

We work the numerator with the relation (1.38) (see 1.8):

$$p(3)_{r=1} = 2(3) + [6(3-2)+1]$$

$1s^2$

$2s^2 \qquad 2p^6$

$3s^2 \qquad 3p^1$

Now the orbital energy of the electron placed in a 3p orbital:

$$E_{3,1} = -\frac{\{s^2 2s^2 2p^6 3s^2 p^1\}^2 \times 13.6 \text{ eV}}{\{3+1)+3(2)\}\{(3)+3(1)\}} = -\frac{\{13\}^2 \times 13.6 \text{ eV}}{2^2 (n=2)}, \quad Z = 13 \text{ (Al)}$$

$$\begin{vmatrix} 0s^1 \\ 1s^1 \\ 2s^1 \quad 2p^3 \\ 3s^1 \quad 3p^3 \end{vmatrix} - \begin{vmatrix} 0s^1 \\ 1s^1 \\ 2s^1 \quad 2p^3 \end{vmatrix} = 3s^1 3p^3 \Rightarrow \text{longitud T}(2) = 2(3s^1)+3p^1 = 3$$

And, once again, the calculation tells us that an electron of H 2p can be placed in a 3p orbital with the energy of the second level despite being brought to the level n = 3. The process continues until completing 6 electrons and ends period 3 with the element Ar, Z = 18:

$$p(3)_{r=6} = 2(3) + [6(3-2)+6] = 18 \ (Ar)$$

$$1s^2$$

$$2s^2 \quad 2p^6$$

$$3s^2 \quad 3p^6$$

Now the orbital energy of the electron placed in a 3p orbital:

$$E_{3,6} = -\frac{\{s^2 2s^2 2p^6 3s^2 p^6\}^2 \times 13.6 \text{ eV}}{\{3+1\}+3(2)\} \cdot \{3\}+3(1)\}} = -\frac{\{18\}^2 \times 13.6 \text{ eV}}{2^2 (n=2)}, \quad Z = 18 \ (Ar)$$

$$\begin{vmatrix} 0s^1 \\ 1s^1 \\ 2s^1 & 2p^3 \\ 3s^1 & 3p^3 \end{vmatrix} - \begin{vmatrix} 0s^1 \\ 1s^1 \\ 2s^1 & 2p^3 \end{vmatrix} = 3s^1 3p^3 \Rightarrow \text{longitud } T(2) = 2(3s^1 3p^3) = 8$$

The period x = 3 is completed with 8 elements. Now we move on to build the level n = 3 for the filling of the p orbitals and the start of a new subshell, the 3d, of maximum angular momentum *l* = 3; that is, s orbitals (*l* = 0), p orbitals (*l* = 1) to d orbitals (*l* = 3) will be filled. However, the orbital function (1.8) must be written in the correct filling sequence, as was done with the element Rb of atomic number Z = 37 (Article 4: "Algebra and Geometry of Periodic Systems of Elements" explains the difference between a system on the right and a system on the left and the correct writing of the functions corresponding to each system).

Evel n = 3 periods x = 4,5   y   r = 1,2,3,4,5 y 6

Next, we will write the function 1.8 in the correct sequence of the elements that are filling the p orbitals:

$$p(x)_{r=1,2,\ldots,6} = 2x + 14(x-5) + 10(x-3) + [6(x-2) + r]$$
(1.39)

The same should be done with the function (1.33) which calculates the number of elements and the type of orbitals in each period of the Periodic Table and the orbital energy function (1.38) is written as follows:

$$E_{x,r} = -\frac{\{2x+14(x-5)+10(x-3)+[6(x-2)+r]\}^2 \times 13.6 \text{ eV}}{\{(x+1)+7(x-5)+5(x-3)+3(x-1)\}\{x+7(x-6)+5(x-4)+3(x-2)\}}$$
(1.40)

To simplify we will calculate for the noble gases Kr (Z =36) and Xe (Z = 54) that close periods 4 and 5, respectively. These elements have filled the p orbitals with r = 6 differential electrons.

Periodo: $x = 4$, $r = 6$ ($l = 2, n = 3$) $Z = 36$ (Kr)

Recall that the relationship (maximum torque-momentum period) is given by $x = 2l \Rightarrow 4 = 2l \Rightarrow l = 2(n=3)$, therefore the energy ratio (1.40) remains

$$E_{x,r} = -\frac{\{2x+10(x-3)+[6(x-2)+r]\}^2 \times 13.6 \text{ eV}}{\{(x+1)+5(x-3)+3(x-1)\}\{x+5(x-4)+3(x-2)\}}$$

We work separately the numerator replacing $x = 4$ and $r = 6$:

$$p(x)_{r=6} = 2x + 10(x-3) + [6(x-2) + r]$$
$$p(4)_{r=6} = 2(4) + 10(4-3) + [6(4-2) + 6] = 36 \, (Kr)$$

$$1s^2$$
$$2s^2 \qquad\qquad 2p^6$$
$$3s^2 \qquad\qquad 3p^6$$
$$4s^2 \quad 3d^{10} \quad 4p^6$$

And is the filling configuration of the element Z = 36. Note that function 1.8 as presented also gives us the increasing order of the configuration in levels and sublevels:

$$p(4)_{r=6} = 2(4) + [6(4-2) + 6] + 10(4-3)$$

$$1s^2$$
$$2s^2 \qquad\qquad 2p^6$$
$$3s^2 \qquad\qquad 3p^6 \qquad\qquad 3d^{10}$$
$$4s^2 \qquad\qquad 4p^6$$

Being, therefore, both valid.

Now we go to the denominator to find that it matches the filling configuration of the element Z = 36 (Kr).

$$E_{4,6} = -\frac{\{2(4) + 10(4-3) + [6(4-2) + 6]\}^2 \times 13.6 \text{ eV}}{\{(4+1) + 5(4-3) + 3(4-1)\} \{4 + 5(4-4) + 3(4-2)\}} = -\frac{\{36\}^2 \times 13.6 \text{ eV}}{3^2 \, (n=3)}$$

$$\begin{vmatrix} 0s^1 \\ 1s^1 \\ 2s^1 \\ 3s^1 \\ 4s^1 \quad 3d^5 \end{vmatrix} \begin{vmatrix} 2p^3 \\ 3p^3 \\ 4p^3 \end{vmatrix} - \begin{vmatrix} 0s^1 \\ 1s^1 \\ 2s^1 \\ 3s^1 \end{vmatrix} \begin{vmatrix} 2p^3 \\ 3p^3 \end{vmatrix} = 2 \times (4s^1 \, 3d^5 \, 4p^3) = 18$$

note, according to this result, that a hydrogen electron 3s, 3p, 3d is going to be placed in the orbital 4s, 3d, 4p with energy n = 3, amazing! Now let's look again at the denominator and we are calculating period by period:

$T(x) = \{(x+1)+5(x-3)+3(x-1)\} - \{x+5(x-4)+3(x-2)\}$

$T(0) = \{(0+1)\} - \{0\} = 1^2 \Rightarrow 0s^1 \Rightarrow 2(0s^1) = 2$ cuadros $x = 0: -1$ y 0 $(n=1)$

$T(1) = \{(1+1)\} - \{1\} = 1^2 \Rightarrow 2(1s^1) = 1s^2 = 2$ cuadrados en $x = 1$: H, He $(n=1)$

$T(2) = \{(2+1)+3(2-1)\} - \{2\} = 2^2 \Rightarrow 2(2s^1 \ 2p^3) = 2s^2 \ 2p^6 = 8 \ (n=2)$

$T(3) = \{(3+1)+3(3-1)\} - \{3+3(1)\} = 2^2 \Rightarrow 2(3s^1 3p^3) = 3s^2 3p^6 = 8 \ (n=2)$

Finalmente

$T(4) = \{(4+1)+5(4-3)+3(4-1)\} - \{4+5(4-4)+3(4-2)\} = 3^2 = 9 \ (n=3)$

$$\begin{vmatrix} 0s^1 \\ 1s^1 \\ 2s^1 & & 2p^3 \\ 3s^1 & & 3p^3 \\ 4s^1 & 3d^5 & 4p^3 \end{vmatrix} - \begin{vmatrix} 0s^1 \\ 1s^1 \\ 2s^1 & 2p^3 \\ 3s^1 & 3p^3 \end{vmatrix} = 2(4s^1 \ 3d^5 \ 4p^3) = 18$$

Note that the electronic configuration obtained with the function p(x) of the numerator, which gives us the atomic number and its distribution, coincides with the function of the denominator T(x), which calculates the length and type of orbitals that are filled in each period, with the exception that the number of electrons per orbital must be assigned; in short, numerator and denominator gives us the configuration:

$$\begin{array}{lll} 1s^2 & & \text{periodo: } x = 1 \\ 2s^2 & 2p^6 & x = 2 \\ 3s^2 & 3p^6 & x = 3 \\ 4s^2 \ 3d^{10} & 4p^6 & x = 4 \end{array}$$

67

We now calculate the orbital energies for the element of atomic number 36 (Kr) and a single electron and whose sequence, although corresponding to a single electron, leads to its correct configuration when two electrons are placed in each corresponding orbital:

Period: $x = 1$, $r = 6$ ($l = 0, n = 1$) $Z = 36$

$$E_{1,6} = -\frac{\{36\}^2 \times 13.6 \text{ eV}}{\{1+1\}\{1\}} = -13.6 \text{ eV} \times (36)^2$$

$$\begin{vmatrix} 0s^1 \\ 1s^1 \end{vmatrix} - |0s^1| = 1s^1 \Rightarrow 2(1s^1) = \boxed{1s^2}$$

Note, to emphasize, that the general calculation refers to a nucleus of Z protons and a single electron that can be placed in the indicated orbitals in each period according to the selected sequence. The electron can be carried to higher levels, but maintaining the energy of the level corresponding to the n of the staggered function of the system

Period: $x = 2$, $r = 6$ ($l = 1, n = 2$) $Z = 36$ (Kr)

$$E_{2,6} = -\frac{\{36\}^2 \times 13.6 \text{ eV}}{\{2+1\}+3(1)\{2\}} = -\frac{\{36\}^2 \times 13.6 \text{ eV}}{2^2 (n=2)}$$

$$\begin{vmatrix} 0s^1 \\ 1s^1 \\ 2s^1 \quad 2p^3 \end{vmatrix} - \begin{vmatrix} 0s^1 \\ 1s^1 \end{vmatrix} = 2s^1 \; 2p^3 \Rightarrow 2(2s^1 \; 2p^3) = \boxed{2s^2 2p^6}$$

Period: $x = 3$, $r = 6$ ($l = 1, n = 2$) $Z = 36$ (Kr)

$$E_{3,1} = -\frac{\{36\}^2 \times 13.6 \text{ eV}}{\{(3+1)+3(2)\}-\{(3)+3(1)\}} = -\frac{\{36\}^2 \times 13.6 \text{ eV}}{2^2 \, (n=2)}$$

$$\begin{vmatrix} 0s^1 \\ 1s^1 \\ 2s^1 & 2p^3 \\ 3s^1 & 3p^3 \end{vmatrix} - \begin{vmatrix} 0s^1 \\ 1s^1 \\ 2s^1 & 2p^3 \end{vmatrix} = 3s^1 3p^3 \Rightarrow 2\left(3s^1 3p^3\right) = \boxed{3s^2 3p^6}$$

Period: $x = 4$, $r = 6$ ($l = 2, n = 3$) $Z = 36$ (Kr)

$$E_{4,6} = -\frac{\{36\}^2 \times 13.6 \text{ eV}}{\{(4+1)+5(4-3)+3(4-1)\}-\{4+5(4-4)+3(4-2)\}} = -\frac{\{36\}^2 \times 13.6 \text{ eV}}{3^2 \, (n=3)}$$

$$\begin{vmatrix} 0s^1 \\ 1s^1 \\ 2s^1 & 2p^3 \\ 3s^1 & 3p^3 \\ 4s^1 & 3d^5 & 4p^3 \end{vmatrix} - \begin{vmatrix} 0s^1 \\ 1s^1 \\ 2s^1 & 2p^3 \\ 3s^1 & 3p^3 \end{vmatrix} = 2 \times \left(4s^1 \, 3d^5 \, 4p^3\right) = \boxed{4s^2 \, 3d^{10} \, 4p^6}$$

Notice, from the calculation made, that in periods 2 and 3 corresponding to n = 2 of the staggered function, figure 1.2, the energy of the electron is the energy of the Atom of H at the level n = 2 multiplied by the atomic number, however, how is it that the electrons are assigned?

§ 8.4 Orbital energies and filling functions for d-block elements

Next, we will write the function 1.10 in the correct sequence of the elements that are filling the d orbitals:

$$d(x)_r = 2x + 14(x-5) + \left[10(x-4)+r\right] + 6(x-2), \quad \forall x \geq 4$$
(1.41)

As explained, the electronic configurations of the elements of the d and f blocks undergo, in appearance, a change that is simply due to the fact that the orbitals that are going to be filled in periods 4, 5, 6 and 7 have the same energy and for greater stability the electron transits occur. However, periodic functions determine exactly the atomic numbers and their configurations by correctly fitting with the experimental results. Consequently, the energy ratio for the filling orbitals of the d-block elements is written as follows: $(x \geq 4)$

$$E_{x,r} = -\frac{\{2x + 14(x-5) + \left[10(x-4)+r\right] + 6(x-2)\}^2 \times 13.6 \text{ eV}}{\{(x+1)+7(x-5)+5(x-3)+3(x-1)\}\{x+7(x-6)+5(x-4)+3(x-2)\}}$$
(1.42)

However, for those elements where electronic transit occurs, the following relationship must be used in the numerator:

$$\begin{aligned}[z]d_r(x) &= \left[2(x-1)+6(x-2)+10(x-4)+14(x-6)\right] \cdot (x-2)f^{14} \\ &\quad (x-1)_d\left[\{x+6(x-2)+\left[10(x-4)+r\right]+14(x-5)\} - \{2x+6(x-2)+10(x-4)+14(x-5)\}\right] + y \\ &\quad \cdot x\,s^{2-y} \qquad \left[r \equiv r = 1,2,3,4,5,6,7,8,9,10. \quad (y = 1 \text{ ó } 2), (x \equiv x)\right]\end{aligned}$$
(1.43)

We have already explained the procedure. Take, for example, the atomic number element found in the Periodic Table in the period $x = 6$ and the column $r = 8$ and has an electronic transit $y = 1$. We apply the relationship (1.43):

$[Z]_{d_8}(6) = [2(6-1)+6(6-2)+10(6-4)] \cdot (6-2)f^{14} \cdot$
$(6-1).d^{[\{2(6)+6(6-2)+[10(6-4)+8]+14(6-5)\} - \{2(6)+6(6-2)+10(6-4)+14(6-5)\}]+1}$
$\cdot 6s^{2-1}$

$[Z]_{d_8}(6) = [2(5)+6(4)+10(2)] \cdot (4)f^{14} \cdot$
$5.d^{[\{12+24+[20+8]+14(1)\} - \{12+24+20+14\}]+1} \, 6s^1$

$[Z]_{d_8}(6) = [10+24+20].(4)f^{14} \cdot (5).d^{[\{78\}-\{70\}]+1} = [Xe]\,4f^{14}\,5d^9\,6s^1$

The sum gives us the atomic number Z = 78, it is the element Pt. The configuration is explicitly determined as follows:

$[Z]_{d_8}(6) = [2(5)+6(4)+10(2)] \cdot 4f^{14} \cdot 5d^9 \cdot 6s^1$

$\begin{vmatrix} 1s^2 & & \\ 2s^2 & 2p^6 & \\ 3s^2 & 3p^6 & 3d^{10} \\ 4s^2 & 4p^6 & 4d^{10} \\ 5s^2 & 5p^6 & \end{vmatrix}$ configuración de Xe, queda

$1s^2 - 2s^2\,2p^6 - 3s^2\,3p^6\,3d^{10} - 4s^2\,4p^6\,4d^{10}\,4f^{14} - 5s^2\,5p^6\,5d^9 - 6s^1$

The configuration has been written in increasing order of the principal quantum number n; sin embargo, it can be written according to the filling sequence or as read in the Periodic Table:

			periodo:	$x = 1$
$1s^2$				
$2s^2$		$2p^6$		$x = 2$
$3s^2$		$3p^6$		$x = 3$
$4s^2$	$3d^{10}$	$4p^6$		$x = 4$
$5s^2$	$4d^{10}$	$5p^6$		$x = 5$
$6s^1$	$4f^{14}$ $5d^9$			$x = 6$

This order of filling is achieved with the expression of the denominator in the relation (1.42) which gives us the energy of the electron in the orbitals to be filled in each period as indicated below:

Period: $x = 1$, $(l = 0, n = 1)$ Z = 78 (Pt)

$$E_{1,8} = -\frac{\{78\}^2 \times 13.6 \text{ eV}}{\{(1+1)\}\cdot\{1\}} = -13.6 \text{ eV} \times (78)^2$$

$$\begin{vmatrix} 0s^1 \\ 1s^1 \end{vmatrix} - |0s^1| = 1s^1 \Rightarrow 2(1s^1) = \boxed{1s^2}$$

Period: $x = 2$, $(l = 1, n = 2)$

$$E_{2,8} = -\frac{\{78\}^2 \times 13.6 \text{ eV}}{\{(2+1)+3(1)\}\cdot\{2\}} = -\frac{\{78\}^2 \times 13.6 \text{ eV}}{2^2 (n=2)}$$

$$\begin{vmatrix} 0s^1 \\ 1s^1 \\ 2s^1 \quad 2p^3 \end{vmatrix} - \begin{vmatrix} 0s^1 \\ 1s^1 \end{vmatrix} = 2s^1 \ 2p^3 \Rightarrow 2(2s^1 \ 2p^3) = \boxed{2s^2 2p^6}$$

Period: $x = 3$, $(l = 1, n = 2)$

$$E_{3,8} = -\frac{\{78\}^2 \times 13.6 \text{ eV}}{\{(3+1)+3(2)\}-\{(3)+3(1)\}} = -\frac{\{78\}^2 \times 13.6 \text{ eV}}{2^2 (n=2)}$$

$$\begin{vmatrix} 0s^1 & & \\ 1s^1 & & \\ 2s^1 & 2p^3 \\ 3s^1 & 3p^3 \end{vmatrix} - \begin{vmatrix} 0s^1 & & \\ 1s^1 & & \\ 2s^1 & 2p^3 \end{vmatrix} = 3s^1 3p^3 \Rightarrow 2(3s^1 3p^3) = \boxed{3s^2 3p^6}$$

Periodo: $x = 4$, $(l = 2, n = 3)$

$$E_{4,8} = -\frac{\{78\}^2 \times 13.6 \text{ eV}}{\{(4+1)+5(4-3)+3(4-1)\}-\{4+5(4-4)+3(4-2)\}} = -\frac{\{78\}^2 \times 13.6 \text{ eV}}{3^2 (n=3)}$$

$$\begin{vmatrix} 0s^1 & & \\ 1s^1 & & \\ 2s^1 & & 2p^3 \\ 3s^1 & & 3p^3 \\ 4s^1 & 3d^5 & 4p^3 \end{vmatrix} - \begin{vmatrix} 0s^1 & & \\ 1s^1 & & \\ 2s^1 & & 2p^3 \\ 3s^1 & & 3p^3 \end{vmatrix} = 2 \times (4s^1 \; 3d^5 \; 4p^3) = \boxed{4s^2 \; 3d^{10} \; 4p^6}$$

Periodo: $x = 5$, $(l = 2, n = 3)$

$$E_{5,8} = -\frac{\{78\}^2 \times 13.6 \text{ eV}}{\{(5+1)+5(5-3)+3(5-1)\}-\{5+5(5-4)+3(5-2)\}} = -\frac{\{78\}^2 \times 13.6 \text{ eV}}{3^2 (n=3)}$$

$$\begin{vmatrix} 0s^1 & & \\ 1s^1 & & \\ 2s^1 & & 2p^3 \\ 3s^1 & & 3p^3 \\ 4s^1 & 3d^5 & 4p^3 \\ 5s^1 & 4d^5 & 5p^3 \end{vmatrix} - \begin{vmatrix} 0s^1 & & \\ 1s^1 & & \\ 2s^1 & & 2p^3 \\ 3s^1 & & 3p^3 \\ 4s^1 & 3d^5 & 4p^3 \end{vmatrix} = 2 \times (5s^1 \; 4d^5 \; 5p^3) = \boxed{5s^2 \; 4d^{10} \; 5p^6}$$

Periodo: $x = 6$, $(l = 3, n = 4)$

$$E_{6,8} = -\frac{\{78\}^2 \times 13.6 \text{ eV}}{\{(6+1)+7(1)+5(3)+3(5)\}\cdot\{6+5(2)+3(4)\}} = -\frac{\{78\}^2 \times 13.6 \text{ eV}}{4^2 \, (n=4)}$$

$$\begin{vmatrix} 0s^1 \\ 1s^1 \\ 2s^1 & & 2p^3 \\ 3s^1 & & 3p^3 \\ 4s^1 & 3d^5 & 4p^3 \\ 5s^1 & 4d^5 & 5p^3 \\ 6s^1 & 4f^7 & 5d^5 & 6p^3 \end{vmatrix} - \begin{vmatrix} 0s^1 \\ 1s^1 \\ 2s^1 & & 2p^3 \\ 3s^1 & & 3p^3 \\ 4s^1 & 3d^5 & 4p^3 \\ 5s^1 & 4d^5 & 5p^3 \end{vmatrix} = 2 \times \boxed{6s^1 \, 4f^7 \, 5d^5 \, 6p^3} = 32 \text{ elementos}$$

The period x = 6 completes its orbitals and length of the period is 32 elements, but the configuration for this element is $6s^1 \, 4f^{14} \, 5d^9$ which gives a path of 24 elements counting from the filling of the orbital 6s and we will see how it is calculated in the next paragraph. Consequently, from period x = 1 to period x = 6 with the path $6s^1 \, 4f^{14} \, 5d^9$ matches the atomic configuration and number:

$1s^2$				periodo: $x = 1$
$2s^2$			$2p^6$	$x = 2$
$3s^2$			$3p^6$	$x = 3$
$4s^2$		$3d^{10}$	$4p^6$	$x = 4$
$5s^2$		$4d^{10}$	$5p^6$	$x = 5$
$6s^1$	$4f^{14}$	$5d^9$		$x = 6$

Note that until period x = 5 you have completely full layers and in period 6 there is a transit of an electron 6s to the 5d orbital since they have the same energy. The behavior of an element both chemically and physically is the result of the periodic repetition of the same configuration of the external electrons and each period of the table begins with a configuration element. s¹ where the journey of each period begins until completely filled layers and ending with an element of noble gas 1s² (He) and the rest with the configuration s² p⁶. We will see the functions of filling orbitals one by one starting the period with an orbital **ns¹** for each state s, p, d and f (see paragraph 9). Now, the interesting thing about this result is that the calculation indicates that an electron of H of energy -13.6/n² eV it can be brought to one of the 6s 4f 5d orbitals and maintain the energy of level n = 4 which corresponds to the n of the step function in Figure 1.2; In addition, we can see that although the calculation of orbital energies 1.42 refers to a nucleus of Z protons and a single electron coincides with the filling of orbitals in each period of the table by placing two electrons in each of the corresponding orbitals. This is a representative example for all the elements of the block of elements that are filling the d orbitals and for each one it fits exactly with the periodic functions that we have developed. Now we will move on to the filling of the f orbitals.

§ 8.5 Orbital energies and filling functions for f-block elements

Next, we will write the function 1.13 in the correct sequence of the elements that are filling the f orbitals:

$$f(x)_r = 2(x-0) + \left[14(x-6)+r\right] + 10(x-4) + 6(x-2), \forall x \geq 6$$

(1.44)

This function is correct for those elements where electronic transit does not occur, such as ytterbium. (Z = 70) (see § 11 the table of electronic configurations for the elements of block f) located in the period x = 6 to which it has filled the 7 orbitals f so that r = 14, with this data we obtain

$$f(x)_r = 2(x-0) + [14(x-6)+r] + 10(x-4) + 6(x-2)$$
$$f(6)_{14} = 2(6-0) + [14(6-6)+14] + 10(6-4) + 6(6-2)$$
$$f(6)_{14} = 2(6) + [14(0)+14] + 10(2) + 6(4) = Z = 70\,(Yb),$$

$$\begin{vmatrix} 1s^2 & & & \\ 2s^2 & & & 2p^6 \\ 3s^2 & & & 3p^6 \\ 4s^2 & & 3d^{10} & 4p^6 \\ 5s^2 & & 4d^{10} & 5p^6 \\ 6s^2 & 4f^{14} & & \end{vmatrix} = 70$$

We have the configuration according to the reading of the Periodic Table or filling sequence; however, we can express the configuration in increasing order of energy levels and their corresponding orbitals:

$$f(6)_{14} = 2(6) + 6(4) + 10(2) + [14(0)+14] = Z = 70\,(Yb),$$

$$\begin{vmatrix} 1s^2 & & & \\ 2s^2 & 2p^6 & & \\ 3s^2 & 3p^6 & 3d^{10} & \\ 4s^2 & 4p^6 & 4d^{10} & 4f^{14} \\ 5s^2 & 5p^6 & & \\ 6s^2 & & & \end{vmatrix}$$

Here we find the double application: 1) we calculate the atomic number of the element according to its location in the table in the period x = 6 and column r = 14 of the block of elements that has filled the orbitals f; 2) its electronic configuration, either according to the reading of the table or the increasing order of the largest main number n. The other function, and which leads to the same result, when there is electronic transit is as follows:

$$[Z]_{f_r}(x) = [2(x-1)+6(x-2)+10(x-4)+14(x-6)] .$$
$$(x-2)_f \{2x+6(x-2)+10(x-4)+[14(x-6)+r]]-[2x+6(x-2)+10(x-4)+14(x-6)]\} y$$
$$. (x-1) d^y . x s^2$$

(1.45)

As already explained, here the first bracket corresponds to the noble gas of fully filled shell located a period before the atomic number element Z of block f in period x. Let's see the application of (1.45) for the element located in the period x = 7 to which the differential electron r = 7 has entered the orbital f but makes an electronic transit y = 1:

$$[Z]_{f_4}(7) = [2(7-1)+6(7-2)+10(7-4)+14(7-6)] .$$
$$(7-2)_f \{2(7)+6(7-2)+10(7-4)+[14(7-6)+4]]-[2(7)+6(7-2)+10(7-4)+14(7-6)]\} 1$$
$$. (7-1) d^1 . 7 s^2$$

$$[Z]_{f_4}(7) = [86] . 5f^{\{92(U)]-[88(Ra)]\}-1} . 6d^1 . 7s^2 = [Rn] . 5f^3 . 6d^1 . 7s^2$$

This result gives exactly the location in the periodic table, Figure 1.3, of the uranium of atomic number Z = 92 in the period x = 7 and the column r = 4 despite passing an electron 5f to the orbital 6d. Note that the electron transition generally occurs in orbitals of the same period and from the bottom up between contiguous orbitals as deduced from Figure 1.2.

Denominator analysis

Always the denominator, for the filling of orbitals in each period and number of elements when assigning the electrons in each type of orbital of the corresponding period, gives the sequence of filling or reading of the periodic system according to the parameter P = R/4, R=1,2,3,4, ... and for the Periodic Table R = 2 electrons per orbital and P = 1/2, which gives us the following filling sequence when completing the 7 periods, x = 7:

$$T(x) = \{(x+1) + 7(x-5) + 5(x-3) + 3(x-1)\} - \{x + 7(x-6) + 5(x-4) + 3(x-2)\}$$
(1.46)

Now we calculate the length of each period and the corresponding orbital(s) starting from the zero period in the Periodic Table:

Period cero: x = 0 (n = 1)

$$T(0) = \{(0+1)\} - \{0\}$$

$$\begin{vmatrix} 0 \\ 0s^1 \end{vmatrix} - |0| = 0s^1 \Rightarrow 2 \times (0s^1) = 0s^2 = 2 \ (0,-1)$$

We note that the exponents of the orbitals will be multiplied by two and the length of each period is found. Note that the interaction of the angular momentum electron wave is being analyzed here $l = 0$ which is the only one that interacts with the nucleus through the zero space between the nucleus and the electron. As indicated, here acts the polarization of the vacuum issue that is beyond the scope of quantum-periodic algebra.

Period: x = 1 (n = 1)

$$T(0) = \{(1+1)\} - \{1\}$$

$$\begin{vmatrix} 0g^1 \\ 1s^1 \end{vmatrix} - |0g^1| = 1s^1 \Rightarrow 2 \times (1s^1) = 1s^2 = 2 \ (H, He)$$

Period: x = 2 (n = 2)

$T(2) = \{(2+1)+3(2-1)\} - \{2\}$

$$2 \times \begin{vmatrix} 0s^1 \\ 1s^1 \\ 2s^1 & 2p^3 \end{vmatrix} - 2 \times \begin{vmatrix} 0s^1 \\ 1s^1 \end{vmatrix} = 2s^2 \ 2p^6 = 8$$

Period: $x = 3$ (n = 2)

$T(3) = \{(3+1)+3(3-1)\} - \{3+3(3-2)\}$

$$2 \times \begin{vmatrix} 0s^1 \\ 1s^1 \\ 2s^1 & 2p^3 \\ 3s^1 & 3p^3 \end{vmatrix} - 2 \times \begin{vmatrix} 0s^1 \\ 1s^1 \\ 2s^1 & 2p^3 \end{vmatrix} = 3s^2 \ 3p^6 = 8$$

Period: $x = 4$ (n = 3)

$T(4) = \{(4+1)+5(4-3)+3(4-1)\} - \{4+3(4-2)\}$

$$2 \times \begin{vmatrix} 0s^1 \\ 1s^1 \\ 2s^1 & & 2p^3 \\ 3s^1 & & 3p^3 \\ 4s^1 & 3d^5 & 4p^3 \end{vmatrix} - 2 \times \begin{vmatrix} 0s^1 \\ 1s^1 \\ 2s^1 & 2p^3 \\ 3s^1 & 3p^3 \end{vmatrix} = 4s^2 \ 3d^{10} \ 4p^6 = 18$$

Period: $x = 5$ (n = 3)

$T(5) = \{(5+1)+5(5-3)+3(5-1)\} - \{5 + 5(5-4) \ 3(5-2)\}$

$$2 \times \begin{vmatrix} 0s^1 \\ 1s^1 \\ 2s^1 & & 2p^3 \\ 3s^1 & & 3p^3 \\ 4s^1 & 3d^5 & 4p^3 \\ 5s^1 & 4d^5 & 5p^3 \end{vmatrix} - 2 \times \begin{vmatrix} 0s^1 \\ 1s^1 \\ 2s^1 & & 2p^3 \\ 3s^1 & 3d^5 & 3p^3 \\ 4s^1 & & 4p^3 \end{vmatrix} = 5s^2 \ 4d^{10} \ 5p^6 = 18$$

Period: $x = 6$ $(n = 4)$

Proceeding in an analogous way, it results in the length in the period $x = 5$ and the type of orbitals to be filled with two electrons each:

$$T(6) = \{(6+1) + 7(6-5) + 5(6-3) + 3(6-1)\} - \{6 + 5(6-4) + 3(6-2)\}$$

$$T(6) = 2 \times \{(6+1) + 7(1) + 5(3) + 3(5)\} - 2 \times \{6 + 5(2) + 3(4)\} = 6s^2 4f^{14} 5d^{10} 6p^6 = 32$$

Note the orbitals to be filled 6s 4f 5d 6p and 32 elements that complete the period $x = 6$.

Period: $x = 7$ $(n = 4)$

$$T(7) = \{(7+1) + 7(7-5) + 5(7-3) + 3(7-1)\} - \{7 + 7(7-6) + 5(7-4) + 3(7-2)\}$$

$$T(7) = 2 \times \{(7+1) + 7(2) + 5(4) + 3(6)\} - 2 \times \{7 + 7(1) + 5(3) + 3(5)\}$$

$$T(7) = 7s^2 \ 5f^{14} \ 6d^{10} \ 7p^6 = 32 \ \text{elementos}$$

We find, therefore, that the periodic length function harmonizes completely with the reading we make of the Periodic Table by constructing the elements in the correct sequence and in each period so that the calculation from the zero period is summarized below:

				Periodo:		
$0s^2$				$x=0$	2	= longitud periódica
$1s^2$				$x=1$	2	
$2s^2$			$2p^6$	$x=2$	8	
$3s^2$			$3p^6$	$x=3$	8	
$4s^2$		$3d^{10}$	$4p^6$	$x=4$	18	
$5s^2$		$4d^{10}$	$5p^6$	$x=5$	18	
$6s^2$	$4f^{14}$	$5d^{10}$	$6p^6$	$x=6$	32	
$7s^2$	$5f^{14}$	$6d^{10}$	$7p^6$	$x=7$	32	

Orbital energies

In general, we explain in paragraph 8.1 how to calculate the energy of an atom of atomic number Z for the elements under the n of the stepped function, Figure 1.2, in a consecutive even and odd period resulting in the same energy of the n step since both expressions are reduced to Bohr's formula (see relation 1.29):

$$E_{x_0} = -\frac{Z^2 13.6}{\frac{[x_0+2]^2}{4}} \text{ eV} \quad \text{si } x_0 = 2l \quad \wedge \quad E_{x_f} = -\frac{Z^2 13.6}{\frac{[x_f+1]^2}{4}} \text{ eV} \quad \text{si } x_f = 2l+1$$

(1.47)

: In addition, these two are summarized in a single one that covers all the periods x = 0, 1, 2, 3, 4, 5, 6, and 7 of the Table, which we rewrite.

$$E_x = -\frac{Z^2 \times 13.6 \text{ eV}}{\left\{(x+1)+\sum_{l=1}^{n-1}(2l+1)[x-(2l-1)]\right\} - \left\{\sum_{l=0}^{n-1}(2l+1)[x-2l]\right\}}$$

(1.48)

Now we go further and the atomic number in (1.47) and (1.48) is replaced by the function that determines the atomic numbers for each block s, p, d and f, explained in paragraphs 8.2 to 8.4; consequently, the orbital energy function for an electron placed in the orbitals to be filled with two electrons in each period calculated with the function (1.46) and coinciding with the elements of the f-block is expressed below:

$$E_{x,r} = -\frac{\left\{2x+[14(x-6)+r]+10(x-4)+6(x-2)\right\}^2 \times 13.6 \text{ eV}}{\left\{(x+1)+7(x-5)+5(x-3)+3(x-1)\right\}-\left\{x+7(x-6)+5(x-4)+3(x-2)\right\}}$$

(1.49)

All these orbital energies are reduced to the Bohr expression by making the corresponding substitutions and reductions.

§ 9 Function that determines the position of the element from the ns¹ orbital at the beginning of the period and for each column of each block s, p, d, f

Each period of the Table of Elements begins with an orbital ns^1 and **n≥0**. **The unpublished** thing is that in the Table there is a zero period. Below are the travel functions or the length of each period as the orbitals are filled according to the electronic moment *l*=0, 1, 2, 3. Each block has its position function until reaching the maximum capacity by adding the elements according to

$$\sum_{l=0}^{n-1} 2(2l+1) = 2(1+3+5+7+\ldots) = 2n^2 \qquad (1.50)$$

For n = 1 (l = 0) we have periods 0 and 1; if n = 2 (l = 1) includes periods 2 and 3; n = 3 (l = 2) periods 4 and 5; up to n = 4 (l = 3) periods 6 and 7. That is, even periods $x_0 = 2l$ and odd periods $x_0 = 2l+1$.

§ 9.1 Position function for s orbitals

Look at Figure 1.2 where the path of each period is vertical, from bottom to top, or its equivalent Figure 1.3 where the path is horizontal, from left to right. For the states s the route is 1(ns¹) hasta 2(ns²). Call Z = 0, 2(He), 10(Ne), 18(Ar), 36(Kr), 54(Xe), 86(Rn), 118(Og) the numbers that close the periods, respectively, x = 0, 1, 2, 3, 4, 5, 6 y 7 . The function for calculating the atomic numbers of the elements of the block that is filling the s orbitals is as follows (see 1.15):

$$s(x)_{r=1,2} = \left[2(x-1)+r\right]+6(x-2)+10(x-4)+14(x-6)+\ldots \quad x \geq 0$$
(1.51)

For example, with the orbital function s in the period x = 2 and r = 1 gives us the atomic number element Z = 3(Li), as follows:

$$s(2)_{r=1} = \left[2(2-1)+1\right] = Z = 3(Li), \text{ y configuración}$$
$$\left|1s^2 \quad 2s^1\right|$$

The element that closes the previous period, x = 1, is the He of atomic number Z = 2. The function for elements that close each period of fully filled shell or that does not support more wave functions is given by (see 1.16):

$$\left[p(x-1)\right] = 2(x-1) + 6(x-2) + 10(x-4) + 14(x-6) \quad (1.52)$$

So, for example, if Li is located in the period x = 3, then He is in the previous period 3 − 1 = 2 and its atomic number and electron configuration is calculated with (1.52):

$$\left[p(2-1)\right] = \left[p(1)\right] = 2(2-1) = Z = 2(He), \text{ de configuración}$$
$$\left|1s^2\right|$$

What's the point? The following: subtract $s(2) - p(1)$:

$$s(2)_{r=1} = \left[2(1)+1\right] - \left[p(1)\right] = 1 \text{ (un cuadrado iniciando } x = 2)$$
$$\left|1s^2 \quad 2s^1\right| \quad \left|1s^2\right| = 2s^1$$

In general, it is true that:

$$T_s(x)_r = \{2(x-1)+r\}+6(x-2)+10(x-4)+14(x-6)\} -$$
$$\{2(x-1)+6(x-2)+10(x-4)+14(x-6)\}$$
(1.53)

We apply (1.53) and calculate the length in each period, including the zero period, at the beginning of each new period with an s orbital:

Period zero: $x = 0$ (n = 1), r = 1, 2

$$T_s(0)_{r=1} = \{2(0-1)+1\} - \{2(0-1)\} = \{-1\} - \{-2\} = 1 \Rightarrow 0s^1 \quad (Z = -1)$$
$$T_s(0)_{r=2} = \{2(0-1)+2\} - \{2(0-1)\} = \{0\} - \{-2\} = 2 \Rightarrow 0s^2 \quad (Z = 0)$$

Note in the Periodic Table in the period zero, x = 0, that the first grid or square has Z = -1 and we subtract Z = -2 (antihelium, Z = -2 if we go to the periodic antimatter system which will be developed in later articles) in the period before zero, that is, x = -1, and it turns out that T(0) = 1 square starting the period zero. For r = 2 of the period zero results T(0) = 2 squares to complete this period.

Period: $x = 1$ (n = 1)

$$T_s(1)_{r=1} = \{2(1-1)+1\} - \{2(1-1)\} = [1(H)] - \{0(Z=0)\} = 1 \Rightarrow 1s^1(H)$$
$$T_s(1)_{r=2} = \{2(1-1)+2\} - \{2(1-1)\} = [2(He)] - \{0(Z=0)\} = 2 \Rightarrow 1s^2(He)$$

Period: $x = 2$ (n = 2)

$$T_s(2)_{r=1} = \{2(2-1)+1\} - \{2(2-1)\} = [3(Li)] - \{2(He)\} = 1 \Rightarrow 2s^1(Li)$$

$$T_s(2)_{r=2} = \{2(2-1)+2\} - \{2(2-1)\} = [4(Be)] - \{2(He)\} = 2 \Rightarrow 2s^2(Be)$$

Period: $x = 3\ (n = 2)$

$$T_s(3)_{r=1} = \{2(3-1)+1]+6(3-2)\}-\{2(3-1)+6(3-2)\}= \{11(Na)]\}-\{10(Ne)\}=3s^1$$

$$T_s(3)_{r=2} = \{2(3-1)+2]+6(3-2)\}-\{2(3-1)+6(3-2)\}= \{12(Mg)]\}-\{10(Ne)\}=3s^2$$

We find, henceforth, that the path of the elements of the Periodic Table always begins with an orbital ns^1 as explained at the beginning of this paragraph. The path of the elements as the orbitals p, d, f are filled in each period is summarized below.

§ 9.2 Position function for the orbitals p si $x \geq 2$

By simple inspection there is the following function:

$$T_p(x)_r = \{2x +[6(x-2)+r]+10(x-3)+14(x-5)\}-$$
$$\{2(x-1)+6(x-2)+10(x-4)+14(x-6)\}$$
(1.54)

As an example, we want to know what the path is in the period $x = 5$ if the differential electron $r = 3$ has entered the orbital p. We do the substitution in this function:

$$T_p(5)_{r=3} = \{2(5)+[6(3)+3]+10(2)\}-\{2(4)+6(3)+10(1)\}=15 \text{ elementos}$$

$$\begin{vmatrix} 1s^2 & & \\ 2s^2 & 2p^6 & \\ 3s^2 & 3p^6 & 3d^{10} \\ 4s^2 & 4p^6 & 4d^{10} \\ 5s^2 & 5p^3 & Sb \end{vmatrix} - \begin{vmatrix} 1s^2 & & \\ 2s^2 & 2p^6 & \\ 3s^2 & 3p^6 & 3d^{10} \\ 4s^2 & 4p^6 & Kr \end{vmatrix} = 5s^2\ 4d^{10}\ 5p^3$$

This is the antimony element, Sb, Z = 51 full-layer Kr, Z = 36.

§ 9.3 Position function for the orbitals d for all $x \geq 4$

The function is as follows:

$$T_d(x)_r = \{2x + 6(x-2) + [10(x-4) + r] + 14(x-5)\} - $$
$$\{2(x-1) + 6(x-2) + 10(x-4) + 14(x-6)\}$$

(1.55)

We make the following observation: some elements of block d and block f make electronic transit and yet the function gives its correct position in the corresponding period and column. As an example, let's calculate the path (position function) of the element located in the period $x = 4$ and the column $r = 9$. We only do the substitution in the function (1.55):

$$T_d(4)_{r=9} = \{(4) + 6(2) + [10(0) + 9]\} - \{2(3) + 6(2)\} = 29 - 18 = 11$$

$$\begin{vmatrix} 1s^2 & & \\ 2s^2 & 2p^6 & \\ 3s^2 & 3p^6 & 3d^9 \\ 4s^2 & & Cu \end{vmatrix} - \begin{vmatrix} 1s^2 & Ar \\ 2s^2 & 2p^6 \\ 3s^2 & 3p^6 \end{vmatrix} = 4s^1 \; 3d^{10}$$

It is the element copper, Cu, Z = 29 of shell Ar, Z = 18. Note that, although electronic transit of the orbital 4s to 3d has occurred, the sum of its exponents still gives 11 elements that is the path of the beginning of the period $x = 4$ until reaching the square of atomic number Z = 29.

§ 9.4 Position function for orbitals f for all $x \geq 6$

The function is as follows:

$$T_f(x)_r = \frac{1}{2}\{(x-0) + 6(x-2) + 10(x-4) + [14(x-6)+r]\} -$$
$$\frac{1}{2}\{(x-1)+6(x-2) + 10(x-4)+14(x-6)\}$$
(1.56)

Let's see the application of this function for an element in the period x = 6 and the column r = 7:

$$T_f(6)_{r=7} = \frac{1}{2}\{(6) + 6(4) + 10(2) + [14(0)+7]\} - \frac{1}{2}\{(5)+6(4) + 10(2)\} = 9$$

$$\begin{vmatrix} 1s^2 & & & \\ 2s^2 & 2p^6 & & \\ 3s^2 & 3p^6 & 3d^{10} & \\ 4s^2 & 4p^6 & 4d^{10} & 4f^7 \\ 5s^2 & 5p^6 & & \\ 6s^2 & & & \\ & & Eu & \end{vmatrix} - \begin{vmatrix} 1s^2 & & & \\ 2s^2 & 2p^6 & & \\ 3s^2 & 3p^6 & 3d^{10} \\ 4s^2 & 4p^6 & 4d^{10} \\ 5s^2 & 5p^6 & Xe \end{vmatrix} = 6s^2\ 4f^7$$

This is the element Europium, Eu, Z = 63 of shell Xe, Z = 54. The path is the ninth element at the beginning of the period x = 6 with the 6s1 orbital up to the 4f⁷ orbital.

§ 10   Results and discussion

Our analysis of paragraphs §1 to §8 fits very well with the Periodic Table and the experimental results. We have demonstrated through very specific examples how functions model the real physical phenomenon which allow to obtain mathematically the atomic numbers of the elements according to the particular block in the column and period corresponding to said block; the number of elements and the type of orbitals that are going to be filled in a period of independent variable x and the functions for the electron configurations of the atoms according to the s, p d and f orbitals being filled.

We can discuss two experimental results that fit very well with the Mathematical Functions of the Periodic Table:

1) Alkali metals are the most electropositive elements since their atoms have the maximum tendency to lose electrons with electronic structure ns¹ y n = 2, 3, 4, 5, 6 y 7 is the number of levels of the atoms in the periods, respectively, $x = 2,3,4,5,6$ y 7 where the alkali metal begins.

Its configuration is, Li: (2)2s¹; Na: (10)3s¹; K: (18)4s¹; Rb: (36)5s¹; Cs: (54)6s¹ y Fr: (86)7s¹. The numbers in parentheses 2, 10, 18, 36, 54, and 86 correspond, respectively, with what atomic numbers of noble gases He, Ne, ar, Kr, Xe and Rn.

The alkali metals are the only ones that when filled at their last level lose it in ionization. Then the ionization energy determined by flame photometry[5] for items below levels n = 2 and n = 3 in the corresponding periods of the system figure 1.2

$$n=2 \begin{cases} x_0 = 2 \\ x = 3 \end{cases} \quad \begin{matrix} Z=3\,(Li) \\ Z=11\,(Na) \end{matrix} \quad \begin{matrix} E_{ionización} = 5.37 \text{ eV} \\ E_{ionización} = 5.12 \text{ eV} \end{matrix}$$

$$n=3 \begin{cases} x_0 = 4 \\ x = 5 \end{cases} \quad \begin{matrix} Z=19\,(K) \\ Z=37\,(Rb) \end{matrix} \quad \begin{matrix} E_{ionización} = 4.32 \text{ eV} \\ E_{ionización} = 4.16 \text{ eV} \end{matrix}$$

The graph in Figure 1.10 shows the energy or ionization potential as a function of the atomic number. We can see that an electron ns¹ has practically the same energy as the electron (n+1)s that is, for example, in the Li (n = 2) when the stable neutral atom is built, the electron enters the level 2s¹ and in the Na (n = 3) the electron enters the level (2 + 1) s¹, but the two electrons have the same energy at different levels although they are below the n = 2 level of the Periodic Table step function, figure 1.2. Extrapolating this experimental result coincides with the quantum-periodic results considered in § 8.1 for a single electron (see example 1.5). Later we will analyze this result in more depth in the following chapters. However, this does not affect the development of the mathematical functions of the Periodic Table analyzed in paragraphs § 1 to § 8. More specifically, if we were to take Figure 1.2 to the orbitals in the hydrogen atom (Z=1) its single electron can be placed in the degenerate orbitals under the n of the step function and be placed, for example, in the 2s¹ and / or 3s¹ subshell and would have the same energy in both subshells $E_2 = -3,4$ eV despite being in the 3s orbital according to (1.29). The same happens with the 2s¹ and 3s¹ orbitals for Li and Na, respectively, in ionization but for Z = 3 and Z = 11.

2) Barrier diagrams and electronic gates, BDEG, fit very well with the experimental results of the electronic configurations of the atoms. As we have seen, the diagrams allow us to visualize without any calculation of the selection rules and symmetry of quantum mechanics where transitions can occur for the elements of blocks d and f. According to the results of the study of the selection and symmetry rules (bibliographic reference 3), it is concluded that **_transitions are allowed only between even and odd states_**, a conclusion that complies with the BDEG (see § 7.3 and § 7.4).

We do not intend here much less to study the quantum model of the atom for Z> 1 electrons whose wave function depends on 3Z parameters. The wave equation to be solved would be mathematically impossible to solve. For example, solving the wave equation for the iron atom of atomic number Z = 26 we will have a wave function of 3x26=78 variables.

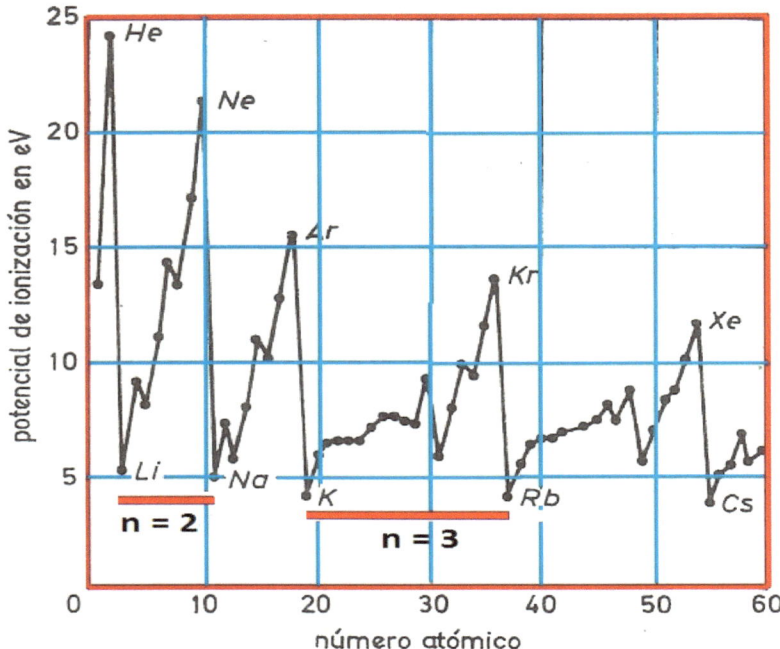

Figure 1.10 Ionization potential of the first electron extracted from the atom as a function of its atomic number. The line for n = 2 is the one that corresponds to the step function for n = 2 in figure 1.2 (left) for the elements Li and Na.   The line for n = 3   for the elements K and Rb in figure 1.2. It is no coincidence that their ionization energies are practically the same.

"A differential equation of seventy-eight variables is obviously impossible[6] to solve" However, the **_quantum-periodic_** result applied to a hydrogen nucleus or hydrogen atoms leads to the beautiful structure of the Periodic Table. So, we still have a long way to go: "in quantum-periodic mechanics no one will ever say the last word; only God"

§ 11 Conclusions and summary of functions and configurations

The new knowledge presented here is only the first chapter of what I have called PERIODIC QUANTUM MECHANICS. This chapter is unique for the development of the **_mathematical functions of the Periodic Table_**, being sufficient for this the quantum results for hydrogen atoms or atoms with nuclei of several protons and a single electron (see introduction). Based on these results and adding the Pauli exclusion principle (not exclusive but periodic quantum algebra supports more than two electrons per orbital) the functions of each block are obtained by counting the atomic Numbers as areas but this procedure is based on a General **_Law of the Periodic Systems_** whose quantum mechanical study is somewhat extensive and encompasses relationships with very varied properties, both algebraic and geometric.

This first part is a "necessary and sufficient condition" without going beyond quantum theory to develop the functions of the Periodic System of Chemical Elements.
In summary:
The Mathematical Construction of the Periodic Table has been systematically developed and its Functions are obtained. The following are summarized the periodic functions corresponding to each block of elements that allow to obtain:

(A) atomic numbers according to period x and column r;

(B) The function for the block B-s, p, d and f that at the same time allow obtaining the electronic configuration without any problem for the blocks that fill the s and p orbitals;

(C) the function to determine the *length in each period* that in turn indicates the *type of subshell that* will fill the orbitals in the corresponding period;

(D) the functions for the electronic configurations for all the blocks according to the type of orbital that the elements fill and that despite certain deviations in the filling of the d and f orbitals the atomic number and its location in the Periodic Table according to column r and period x continue to be correct.

This deviation is explained by the "Barrier Diagrams and Electronic Gates", which indicates without complicated calculations of quantum mechanics the transit of y = 1,2 electrons that agree with the experiment;

(E) Briefly, the fan is opened from *the relationship that determines the energy levels of hydrogen atoms*, introducing the filling sequence that leads to the Periodic Table.

Mathematical functions of the Periodic Table for each block:

Block s: $n = 1 (l = 0)$ $\left[s_r(x) = z \right], r = 1,2$

$$s(x)_r = \left[2(x-1) + r \right] + 6(x-2) + 10(x-4) + 14(x-6) \quad \forall x \geq 0$$

$$[z]_{s(x)_r} = \left[p(x-1) \right] x \; s^{z - \left[2(x-1) + 6(x-2) + 10(x-4) + 14(x-6) \right]}$$

Block p: $n = 2 \,(l=1)\; [p_r(x) = z],\; r = 1,2,3,4,5,6$

$$p(x)_r = 2x + \Big[6(x-2) + r\Big] + 10(x-3) + 14(x-5)\; \forall x \geq 2$$

$$[z]p_r(x) = \Big[p(x-1)\Big] \cdot (x-2)f^{14} \cdot (x-1)d^{10} \cdot x\, s^2 \cdot x\, p^{z - d_{10}(x)}$$

Block d: $n = 3 \,(l=2)\; [d_r(x) = z],\; r = 1, 2, 3, \ldots, 10$

$$d(x)_r = 2x + 6(x-2) + \Big[10(x-4) + r\Big] + 14(x-5)\; \forall x \geq 4$$

$$[z]d_r(x) = \Big[p(x-1)\Big] \cdot (x-2)f^{14} \cdot (x-1)d^{\big[z - (f_{14}(x))\big] + y} \cdot x\, s^{2-y}$$

Block f: $n = 4 \,(l=3)\; [f_r(x) = z],\; r = 1, 2, 3, 4, \ldots, 14$

$$f(x)_r = 2(x-0) + 6(x-2) + 10(x-4) + \Big[14(x-6) + r\Big],\; \forall x \geq 6$$

$$[z]f_r(x) = \Big[p(x-1)\Big](x-2)f^{\big[z - g_{18}(x)\big] - y} (x-1)d^y\, x\, s^2$$

Function for periodic length

$$T(x) = \left\{ 2(x+1) + \sum_{l=1}^{n-1} 2(2l+1)\left[x - (2l-1)\right] \right\}$$
$$- \left\{ 2x + \sum_{l=1}^{n-1} 2(2l+1)\left[x - 2l\right] \right\}$$

This function calculates not only the length or number of elements in each period of the Periodic Table but also the type of orbitals that will be filled in the correct sequence of each period. The Periodic Table only manages to fill up to the f-orbitals of angular momentum $l = 2, n = 3$.

It is satisfied so $x \geq 0$ that for the Periodic Table it allows to calculate two "Tables" with the codes -1 and 0 in the period $x = 0$ (see figure 1.2 y/o figure 1.3). Note that the function $T(x)$ covers $x = 7$ periods in the Periodic Table but the summation develops only up to the quantum number of energy $n = 4$ for the angular momentum $l = 3$ which corresponds to the states f that will be filled with electrons. This function not only calculates the "frames" in each period but also the type of orbitals of the sublayers that are filled with two electrons from the period $x = 1$.

Next, the above functions are written in a long form and the electronic configurations of the elements are calculated not in the filling sequence but in order of levels and sublevels as the atoms finally remain.

The Electron Configurations of the Elements: Long electronic functions

For each block of elements according to the s, p, d and f orbital that has been filled with electrons, the preceding electronic functions have been written in full, resulting in a long form. This allows to know the atomic number of the element and its configuration just by substituting the period x (number of layers) and the column r (differential electron) of the corresponding s, p, d and f block in the Periodic Table. The elements that apparently suffer a deviation in their configuration (blocks d and f) are indicated with an asterisk (*), indicating y = 1 or 2 electrons out of phase. The symbol or atomic number of the noble gases whose function is

$[2(x-1)+6(x-2)+10(x-4)+14(x-6)]$ including zero element $[0]$, to indicate the core of the atom corresponding to stable spherical symmetry layers and sublayers ns^2np^6 $\forall n \geq 2$ [Ne, Ar, Kr, Xe y Rn].

ELEMENTS of the s ORBITALS BLOCK: $n = 1$ $(l=0)$ $\forall x \geq 0$

$$[z]_{s_r}(x) = [2(x-1)+6(x-2)+10(x-4)+14(x-6)] \cdot$$

$$\begin{array}{c} \{2(x-1)+r] +6(x-2)+10(x-4)+14(x-6)+...\} - \\ x \; s \qquad\qquad [2(x-1) + 6(x-2) + 10(x-4) + 14(x-6) + ...] \end{array}$$

r	x	Z	Elements	Configuration	r	x	Z	Elements	Configuration
1	1	1	H	$[0].1s^1$	2	1	2	He	$[0].1s^2$
1	2	3	Li	$[He].2s^1$	2	2	4	Be	$[He].2s^2$
1	3	11	Na	$[Ne].3s^1$	2	3	12	Mg	$[Ne].3s^2$
1	4	19	K	$[Ar].4s^1$	2	4	20	Ca	$[Ar].4s^2$
1	5	37	Rb	$[Kr].5s^1$	2	5	38	Sr	$[Kr].5s^2$
1	6	55	Cs	$[Xe].6s^1$	2	6	56	Ba	$[Xe].6s^2$
1	7	87	Fr	$[Rn].7s^1$	2	7	88	Ra	$[Rn].7s^2$

ELEMENTS of the p ORBITALS BLOCK: $n = 2$ $(l=1)$ $\forall x \geq 2$

$$CE : [Z]_{p_r}(x) = [2(x-1)+6(x-2)+10(x-4)+14(x-6)].(x-2)f^{14}.(x-1)d^{10}.xs^2$$

$$x\,p\{2x+[6(x-2)+r]+10(x-3)+14(x-5)\} - [2x+6(x-2)+10(x-3)+14(x-5)+\ldots]$$

$r = 1,2,3,4,5$ and 6

r	x	Z	Elements	Configuration	r	x	Z	Elements	Configuration
1	2	5	B	$[He]2s^2 2p^1$	2	2	6	C	$[He]2s^2 2p^2$
1	3	13	Al	$[Ne]3s^2 3p^1$	2	3	14	Si	$[Ne]3s^2 3p^2$
1	4	31	Ga	$[Ar]3d^{10} 4s^2 4p^1$	2	4	32	Ge	$[Ar]3d^{10} 4s^2 4p^2$
1	5	49	In	$[Kr]4d^{10} 5s^2 5p^1$	2	5	50	Sn	$[Kr]4d^{10} 5s^2 5p^2$
1	6	81	Tl	$[Xe]4f^{14} 5d^{10} 6s^2 6p^1$	2	6	82	Pb	$[Xe]4f^{14} 5d^{10} 6s^2 6p^2$
1	7	113		$[Rn]5f^{14} 6d^{10} 7s^2 7p^1$	2	7	114	Pb	$[Rn]5f^{14} 6d^{10} 7s^2 7p^2$

r	x	Z	Elements	Configuration	r	x	Z	Elements	Configuration
3	2	7	N	$[He]2s^2 2p^3$	4	2	8	O	$[He]2s^2 2p^4$
3	3	15	P	$[Ne]3s^2 3p^3$	4	3	16	S	$[Ne]3s^2 3p^4$
3	4	33	As	$[Ar]3d^{10} 4s^2 4p^3$	4	4	34	Se	$[Ar]3d^{10} 4s^2 4p^4$
3	5	51	Sb	$[Kr]4d^{10} 5s^2 5p^1$	4	5	52	Te	$[Kr]4d^{10} 5s^2 5p^4$
3	6	83	Tl	$[Xe]4f^{14} 5d^{10} 6s^2 6p^3$	4	6	84	Po	$[Xe]4f^{14} 5d^{10} 6s^2 6p^4$
1	7	115		$[Rn]5f^{14} 6d^{10} 7s^2 7p^3$	4	7	116		$[Rn]5f^{14} 6d^{10} 7s^2 7p^4$

r	x	Z	Elements	Configuration	r	x	Z	Elements	Configuration
5	2	9	F	$[He]2s^2 2p^5$	6	2	10	Ne	$[He]2s^2 2p^6$
5	3	17	Cl	$[Ne]3s^2 3p^5$	6	3	18	Ar	$[Ne]3s^2 3p^6$
5	4	35	Br	$[Ar]3d^{10} 4s^2 4p^5$	6	4	36	Kr	$[Ar]3d^{10} 4s^2 4p^6$
5	5	53	I	$[Kr]4d^{10} 5s^2 5p^5$	6	5	54	Xe	$[Kr]4d^{10} 5s^2 5p^6$
5	6	85	At	$[Xe]4f^{14} 5d^{10} 6s^2 6p^5$	6	6	86	Rn	$[Xe]4f^{14} 5d^{10} 6s^2 6p^6$
5	7	117		$[Rn]5f^{14} 6d^{10} 7s^2 7p^5$	6	7	118		$[Rn]5f^{14} 6d^{10} 7s^2 7p^6$

ELEMENTS of the d ORBITALS BLOCK: $n = 3$ $(l=2)$ $\forall x \geq 4$

$$CE: [Z]_{d_r}(x) = [2(x-1)+6(x-2)+10(x-4)+14(x-6)] \cdot (x-2) f^{14} \cdot$$

$$(x-1) \cdot d^{\left[\{2x+6(x-2)+[10(x-4)+r]+14(x-5)\} - \{2x+6(x-2)+10(x-4)+14(x-5)\}\right] + y}$$

$$\cdot x \, s^{2-y} \qquad [r \equiv r = 1,2,3,4,5,6,7,8,9,10. \quad (y = 1 \text{ ó } 2), (x \equiv x)]$$

Elements that experimentally and apparently deviate from their configuration in the Periodic Table are found in blocks d and f. However, the functions for these blocks give exactly their position in the table in the corresponding column and period. Furthermore, as explained in the DBPEs, functions are constructed for the electronic configuration as determined by the experiment and indicates the electronic transit.

r	x	Z	Elements	Configuration	r	x	Z	Elements	Configuration
1	4	21	Sc	$[Ar]3d^1 4s^2$	2	4	22	Ti	$[Ar]3d^2 4s^2$
1	5	39	Y	$[Kr]4d^1 5s^2$	2	5	40	Zr	$[Kr]4d^2 5s^2$
1	6	71	Lu	$[Xe]4f^{14} 5d^1 6s^2$	2	6	72	Hf	$[Xe]4f^{14} 5d^2 6s^2$
1	7	103	Lw	$[Rn]5f^{14} 6d^1 7s^2$	2	7	104	Unq	$[Rn]5f^{14} 6d^2 7s^2$

r	x	Z	Elements	Configuration	r	x	Z	Elements	Configuration
3	4	23	V	$[Ar]3d^3 4s^2$	4	4	24	*Cr(y=1)	$[Ar]3d^5 4s^1$
3	5	41	*Nb(y=1)	$[Kr]4d^4 5s^1$	4	5	42	*Mo(y=1)	$[Kr]4d^5 5s^1$
3	6	73	Ta	$[Xe]4f^{14} 5d^3 6s^2$	4	6	74	W	$[Xe]4f^{14} 5d^4 6s^2$
3	7	105	Unp	$[Rn]5f^{14} 6d^3 7s^2$	4	7	106	Unh	$[Rn]5f^{14} 6d^4 7s^2$

r	x	Z	Elements	Configuration	r	x	Z	Elements	Configuration
5	4	25	Mn	$[Ar]3d^5 4s^2$	6	4	26	Fe	$[Ar]3d^6 4s^2$
5	5	43	Tc	$[Kr]4d^5 5s^2$	6	5	44	*Ru(y=1)	$[Kr]4d^7 5s^1$
5	6	75	Re	$[Xe]4f^{14} 5d^5 6s^2$	6	6	76	Os	$[Xe]4f^{14} 5d^6 6s^2$
5	7	107	Uns	$[Rn]5f^{14} 6d^5 7s^2$	6	7	108	Uno	$[Rn]5f^{14} 6d^6 7s^2$

r	x	Z	Elements	Configuration	r	x	Z	Elements	Configuration
7	4	27	Co	$[Ar]3d^7 4s^2$	8	4	28	Ni	$[Ar]3d^8 4s^2$
7	5	45	*Rh(y=1)	$[Kr]4d^8 5s^1$	8	5	46	*Pd(y=2)	$[Kr]4d^{10} 5s^0$
7	6	77	Ir	$[Xe]4f^{14} 5d^7 6s^2$	8	6	78	*Pt(y=1)	$[Xe]4f^{14} 5d^9 6s^1$
7	7	109	Une	$[Rn]5f^{14} 6d^7 7s^2$	8	7	110	Unn	$[Rn]5f^{14} 6d^8 7s^2$

r	x	Z	Elements	Configuration	r	x	Z	Elements	Configuration
9	4	29	*Cu(y=1)	$[Ar]3d^{10}4s^1$	10	4	30	Zn	$[Ar]3d^{10}4s^2$
9	5	47	*Ag(y=1)	$[Kr]4d^{10}5s^1$	10	5	48	Cd	$[Kr]4d^{10}5s^2$
9	6	79	*Au(y=1)	$[Xe]4f^{14}5d^{10}6s^1$	10	6	80	Hg	$[Xe]4f^{14}5d^{10}6s^2$
9	7	111	Une	$[Rn]5f^{14}6d^97s^2$	10	7	112	Unb	$[Rn]5f^{14}6d^{10}7s^2$

ELEMENTS of the f ORBITALS BLOCK: $n = 4\ (l = 3)\ \forall x \geq 6$

$CE: \quad [Z]_{f_r}(x) = [2(x-1)+6(x-2)+10(x-4)+14(x-6)]$.

$(x-2)_f\{[2x+6(x-2)+10(x-4)+[14(x-6)+r]]-[2x+6(x-2)+10(x-4)+14(x-6)]\} \cdot y$

$\cdot (x-1)d^y \cdot x\ s^2$

$[Z]f_r(x)$: Read electron configuration of the element of atomic number Z of the Block-f in the column $1 \leq r \leq 14$ of x layers (period x in the Periodic Table).

r	x	Z	Elements	Configuration	r	x	Z	Elements	Configuration
1	6	57	*La(y=1)	$[Xe]5d^16s^2$	2	6	58	*Ce(y=1)	$[Xe]4f^15d^16s^2$
1	7	89	*Ac(y=1)	$[Rn]6d^17s^2$	2	7	90	*Th(y=2)	$[Rn]5f^06d^27s^2$

r	x	Z	Elements	Configuration	r	x	Z	Elements	Configuration
3	6	59	Pr	$[Xe]4f^36s^2$	4	6	60	Nd	$[Xe]4f^45d^06s^2$
3	7	91	*Pa(y=1)	$[Rn]5f^26d^17s^2$	4	7	92	*U(y=1)	$[Rn]5f^36d^17s^2$

r	x	Z	Elements	Configuration	r	x	Z	Elements	Configuration
5	6	61	Pm	$[Xe]4f^5 6s^2$	6	6	62	Sm	$[Xe]4f^6 5d^0 6s^2$
5	7	93	*Np(y=1)	$[Rn]5f^4 6d^1 7s^2$	6	7	94	Pu	$[Rn]5f^6 6d^0 7s^2$

r	x	Z	Elements	Configuration	r	x	Z	Elements	Configuration
7	6	63	Eu	$[Xe]4f^7 6s^2$	8	6	64	*Gd(y=1)	$[Xe]4f^7 5d^1 6s^2$
7	7	95	Am	$[Rn]5f^7 6d^0 7s^2$	8	7	96	*Cm(y=1)	$[Rn]5f^7 6d^1 7s^2$

r	x	Z	Elements	Configuration	r	x	Z	Elements	Configuration
9	6	65	Tb	$[Xe]4f^9 6s^2$	10	6	66	Dy	$[Xe]4f^{10} 5d^0 6s^2$
9	7	97	Bk	$[Rn]5f^9 6d^0 7s^2$	10	7	98	Cf	$[Rn]5f^{10} 6d^0 7s^2$

r	x	Z	Elements	Configuration	r	x	Z	Elements	Configuration
11	6	67	Ho	$[Xe]4f^{11} 6s^2$	12	6	68	Er	$[Xe]4f^{12} 5d^0 6s^2$
11	7	99	Es	$[Rn]5f^{11} 6d^0 7s^2$	12	7	100	Fm	$[Rn]5f^{12} 6d^0 7s^2$

r	x	Z	Elements	Configuration	r	x	Z	Elements	Configuration
13	6	69	Tm	$[Xe]4f^{13} 6s^2$	14	6	70	Yb	$[Xe]4f^{14} 5d^0 6s^2$
13	7	101	Md	$[Rn]5f^{13} 6d^0 7s^2$	14	7	102	No	$[Rn]5f^{14} 6d^0 7s^2$

appendix

103 A: Deduction from the wave equation of hydrogen atoms of the ratio that determines energy levels (1.27): $Za^{-1} - c - cl = 0$ [1]

110 B: Resolution of Radial equation A.2: zero energy level and n ≥ 1 levels from the quantum-periodic fundamental relationship

117 C. Obtained roots (B.15) are substituted in (B.17) for energy levels

119 D. New _periodic quantum energy relationships

124 E. How the periodic hydrogen and periodic table systems of the denominator of energy relations are constructed

134 F. Calculation of the general term of the succession formed by the number of orbitals of the hydrogen system in the design of the Periodic Table

A: Deduction from the wave equation of hydrogen atoms of the ratio that determines energy levels (1.27): $Za^{-1} - c - cl = 0$ [1]

The wave equation (A.1) was proposed in 1926 (close to 100 years) by Edwin Schroedinger in 4 articles in which he developed a new formalism that he applied, among others, to the hydrogen atom reaching the Bohr relation $E_n = -\dfrac{Z^2 \times 13.6\ eV}{n^2}$, who obtained it in 1913, although Schroedinger reached the form (1.26) (see A.5):

$$\frac{1}{r^2}\frac{\partial}{\partial r}\left(r^2 \frac{\partial}{\partial r} f\right) + \frac{1}{r^2 sen\theta}\frac{\partial}{\partial \theta}\left(sen\theta \frac{\partial}{\partial \theta} f\right) + \frac{1}{r^2 sen^2\theta}\frac{\partial^2}{\partial \phi^2}f + \frac{2u}{\hbar^2}\left[E - V(r)\right]f = 0$$

(A.1)

The mathematical procedure of this PARTIAL DIFFERENTIAL EQUATION is well documented (see the bibliography, especially 1 and 4 which we will follow and simplify). We are interested in the results and, most importantly, where in the solution it results (1.27). Here is a summary:

1.- A.1 admits separation of variables: $f(r, \theta, \phi) = R(r)P(\theta)Q(\phi)$

2. Separation of variables leads to three ordinary equations, one for each variable $R(r)$, $P(\theta)$ y $Q(\phi)$. Then each equation and its solution, then a table of values for each radial wave function and the separate angular part.

First equation for the variable r

$$\frac{1}{r^2}\frac{d}{dr}\left(r^2\frac{d}{dr}R(r)\right) + \left[\frac{2u}{\hbar^2}(E-V(r)) - \frac{l(l+1)}{r^2}\right]R(r) = 0 \quad \text{(A.2)}$$

It has as a solution for the radial part of this equation the function

$$R_{nl}(r) = -\left[\left(\frac{2Z}{na_0}\right)^3 \frac{(n-l-1)!}{2n\left[(n+l)!\right]^3}\right]^{\frac{1}{2}} e^{-\frac{\rho}{2}} \cdot \rho^l \, L_{n+l}^{2l+1}(\rho)$$

$$\rho = \frac{2Z}{na_0}r$$

(A.3)

The associated polynomials of Laguerre $L_{n+l}^{2l+1}(\rho)$ are defined by

$$L_{n+l}^{2l+1}(\rho) = \frac{d^{2l+1}}{d\rho^{2l+1}}\left[e^\rho \frac{d^{n+l}}{d\rho^{n+l}}\left(\rho^{n+l}\cdot e^{-\rho}\right)\right] \quad (\rho=rho) \quad \text{(A.4)}$$

And we can calculate the radial functions

$R_{10}(r)$, $R_{20}(r)$, $R_{21}(r)$,... and some functions, tables 1 and 2. Eigenvalues of energy (note that Eq. (A.2) contains E) are given by

$$E_n = -Z^2 \frac{13.6 \text{ eV}}{(n_r + l + 1)^2} = -Z^2 \frac{13.6 \text{ eV}}{n^2}, \quad n=1,2,3,\ldots \quad \text{(A.5)}$$

Second equation for the variable θ:

$$\boxed{\frac{1}{\text{sen}\theta}\frac{d}{d\theta}\left(\text{sen}\theta\frac{dR(\theta)}{d\theta}\right) - \frac{m^2 R(\theta)}{\text{sen}^2\theta} + LR(\theta) = 0} \quad \text{(A.6)}$$

The solution of this differential equation normalized N to unity and which gives the factor theta θ of the eigenfunctions of angular momentum (Another notation used is $\Theta_{lm}(\theta) \equiv P_{lm}(\theta)$)

$$P_{lm}(\theta)_N = \sqrt{\frac{(2l+1)(l-|m|)}{2(l+|m|)}} P_{lm}(\cos\theta) \quad \text{(A.7)}$$

Where the associated functions of Legendre correspond to $P_{lm}(\cos\theta)$

$$P_{lm}(w) = \frac{1}{2^l(l)!}(1-w^2)^{\frac{|m|}{2}} \frac{d^{|m|+l}}{dw^{|m|+l}}(w^2-1)^l \quad \text{(A.8)}$$

Table 3 gives some calculated values for that correspond to the factor $P_{lm}(\theta)$ theta² of the functions of the electronic angular momentum. And eigenvalues:

Eigenvalues of orbital angular momentum for the hydrogen atom:

$$L = \hbar\sqrt{l(l+1)} = \hbar\sqrt{\mathbf{l}(\mathbf{l}+1)} \Rightarrow l \equiv \mathbf{l} = 0,1,2,3,... \qquad \text{(A.9)}$$

Note: notation used interchangeably for angular momentum is:

$$l \equiv \mathbf{l} \equiv \mathbf{1}\,(\mathbf{l} \neq \mathbf{1})$$

Third equation for the variable $\varphi \equiv \phi$

$$\frac{d^2 Q(\phi)}{d\phi^2} + m^2 Q(\phi) = 0 \qquad \text{(A.10)}$$

It has as a solution already standardized

$$\boxed{Q(\phi) = \frac{1}{\sqrt{2\pi}} e^{im\phi}} \qquad \text{(A.11)}$$

And the eigenvalues for the angular momentum component L_z of the electron in the H atom are quantized according to m:

$$L_z = m\hbar, \quad \forall m = 0, \pm 1, \pm 2, \pm 3,... \text{ ó } m = -l, (-l+1),...,0,...,+(l-1), +l \qquad \text{(A.12)}$$

Note: These are the functions corresponding to the hydrogen system. To move to the periodic table, the quantum number of angular momentum l = 0,1,2,3,...,n-1 is replaced in (1.27) by:

$l = \alpha_0 / 2$ (periodos iniciales) y $l = (\alpha_f - 1)/2$ (periodos finales)

See section 8.1, where the calculation and its application are explained.

Orbital wave functions

This function is the product of three functions, each depends on a single coordinate and is expressed as follows:

$$f_{n\,l\,m}(r,\theta,\phi) = R_{nl}(r) P_{lm}(\theta) Q_m(\phi)$$

Each function in this equation is as follows:

Table 1   Associated Laguerre polynomials

$n=1 \Rightarrow \ell=0$	(1s)	$L_1^1(x) = -1!$
$n=2 \Rightarrow \ell=0$	(2s)	$L_2^1(x) = -2!(2-x)$
$\ell=1$	(2p)	$L_3^3(x) = -3!$
$n=3 \Rightarrow \ell=0$	(3s)	$L_3^1(x) = -3!\left(3 - 3x + \dfrac{x^2}{2}\right)$
$\ell=1$	(3p)	$L_4^3(x) = -4!(4-x)$
$\ell=2$	(3d)	$L_5^5(x) = -5!$
$n=4 \Rightarrow \ell=0$	(4s)	$L_4^1(x) = -4!\left(4 - 6x + 2x^2 - \dfrac{x^3}{6}\right)$
$\ell=1$	(4p)	$L_5^3(x) = -5!\left(10 - 5x + \dfrac{x^2}{2}\right)$
$\ell=2$	(4d)	$L_6^5(x) = -6!(6-x)$
$\ell=3$	(4f)	$L_7^7(x) = -7!$

$$\left(x = \rho = \dfrac{2Zr}{na_0}\right)$$

Radial factors $R_{n\ell}(r)$ of wave functions $f_{n\ell m}(r,\theta,\phi)$

Table 2[2,3]

$n=1 \Rightarrow \ell=0$	(1s)	$R_{1s}(r) = \left(\dfrac{Z}{a}\right)^{3/2} 2e^{-\rho/2}$	$\rho = \dfrac{2Z}{na}r$
$n=2 \Rightarrow \ell=0$	(2s)	$R_{2s}(r) = \dfrac{1}{2\sqrt{2}}\left(\dfrac{Z}{a}\right)^{3/2}(2-\rho)e^{-\rho/2}$	
$\ell=1$	(2p)	$R_{2p}(r) = \dfrac{1}{2\sqrt{6}}\left(\dfrac{Z}{a}\right)^{3/2}\rho e^{-\rho/2}$	
$n=3 \Rightarrow \ell=0$	(3s)	$R_{3s}(r) = \dfrac{1}{9\sqrt{3}}\left(\dfrac{Z}{a}\right)^{3/2}(6-6\rho+\rho^2)e^{-\rho/2}$	
$\ell=1$	(3p)	$R_{3p}(r) = \dfrac{1}{9\sqrt{6}}\left(\dfrac{Z}{a}\right)^{3/2}(4-\rho)\rho e^{-\rho/2}$	
$\ell=2$	(3d)	$R_{3d}(r) = \dfrac{1}{9\sqrt{30}}\left(\dfrac{Z}{a}\right)^{3/2}\rho^2 e^{-\rho/2}$	
$n=4 \Rightarrow \ell=0$	(4s)	$R_{4s}(r) = \dfrac{1}{96}\left(\dfrac{Z}{a}\right)^{3/2}(24-36\rho+12\rho^2-\rho^3)e^{-\rho/2}$	
$\ell=1$	(4p)	$R_{4p}(r) = \dfrac{1}{32\sqrt{15}}\left(\dfrac{Z}{a}\right)^{3/2}(20-10\rho+\rho^2)\rho e^{-\rho/2}$	
$\ell=2$	(4d)	$R_{4d}(r) = \dfrac{1}{96\sqrt{5}}\left(\dfrac{Z}{a}\right)^{3/2}(6-\rho)\rho^2 e^{-\rho/2}$	
$\ell=3$	(4f)	$R_{4f}(r) = \dfrac{1}{93\sqrt{35}}\left(\dfrac{Z}{a}\right)^{3/2}\rho^3 e^{-\rho/2}$	

Table 3

$l=0$, orbitales s	$l=1$, orbitales p
$P_{0,0}(\theta) = \dfrac{\sqrt{2}}{2}$	$P_{1,0}(\theta) = \dfrac{\sqrt{6}}{2}\cos\theta$
	$P_{1,\pm 1}(\theta) = \dfrac{\sqrt{3}}{2}\sin\theta$

$l=2$, orbitales d	$l=3$, orbitales f
$P_{2,0}(\theta) = \dfrac{\sqrt{10}}{4}\left(3\cos^2\theta - 1\right)$	$P_{3,0}(\theta) = \dfrac{3\sqrt{14}}{4}\left(\dfrac{5}{3}\cos^3\theta - \cos\theta\right)$
$P_{2,\pm 1}(\theta) = \dfrac{\sqrt{15}}{2}\sin\theta\cos\theta$	$P_{3,\pm 1}(\theta) = \dfrac{\sqrt{42}}{8}\sin\theta\left(5\cos^2\theta - 1\right)$
$P_{2,\pm 2}(\theta) = \dfrac{\sqrt{15}}{4}\sin^2\theta$	$P_{3,\pm 2}(\theta) = \dfrac{\sqrt{105}}{4}\sin^2\theta\cos\theta \qquad P_{3,\pm 3}(\theta) = \dfrac{\sqrt{70}}{8}\sin^2\theta$

Table 4[2,3]

$Q_m(\phi) \equiv \Phi_m(\phi)$ Complejo			$Q_m(\phi)$		Real
$Q_0(\phi)$	=	$\left(1/\sqrt{2\pi}\right)$	$Q_0(\phi)$	=	$\left(1/\sqrt{2\pi}\right)$
$Q_1(\phi)$	=	$\left(1/\sqrt{2\pi}\right)e^{+i\phi}$	$Q_{1\cos}(\phi)$	=	$\left(1/\sqrt{\pi}\right)\cos\phi$
$Q_{-1}(\phi)$	=	$\left(1/\sqrt{2\pi}\right)e^{-i\phi}$	$Q_{1\sin}(\phi)$	=	$\left(1/\sqrt{\pi}\right)\text{Sen}\,\phi$
$Q_2(\phi)$	=	$\left(1/\sqrt{2\pi}\right)e^{+2i\phi}$	$Q_{2\cos}(\phi)$	=	$\left(1/\sqrt{\pi}\right)\cos 2\phi$
$Q_{-2}(\phi)$	=	$\left(1/\sqrt{2\pi}\right)e^{-2i\phi}$	$Q_{2\sin}(\phi)$	=	$\left(1/\sqrt{\pi}\right)\sin 2\phi$

B: Resolution of Radial equation A.2: zero energy level and n ≥ 1 levels from the quantum-periodic fundamental relationship

The procedure is well documented in bibliographic references 1 and 4. The purpose is to arrive at the unpublished relationship (1.27) and its application to the Periodic Table and systems in general. Consequently, we will summarize the steps until we reach (1.27) following the calculation tools:

1) in A.2 we develop the first term and call it $a = \hbar^2 / ue^2$ result

$$R'' + \frac{2}{r} R' + \left[\frac{2E}{a e^2} + \frac{2Z}{a r} - \frac{l(l+1)}{r^2} \right] R = 0$$

(B.1)

The behavior of the radial equation (B.1) at large and short distances is studied:

For large values of r $(r \to \infty)$ the equation (B.1) is

$$R'' - c^2 R = 0 \quad \wedge \quad c^2 = -\frac{2E}{ae^2}$$

This equation has as a solution for large values of r

$$R(r): e^{\pm c.r} \quad \wedge \quad c = \pm\sqrt{\frac{-2E}{a.e^2}} \qquad \text{(B.2)}$$

Wave function when r tends to zero $(r \to 0)$

For this we multiply e^{-cr} by another function K(r) in power series such that the radial function R(r) is limited both for $r \to \infty$ as for $r \to 0$.

$$R(r) = e^{-cr} K(r) \quad (e = \text{base del logaritmo natural}) \quad \textbf{(B.3)}$$

From R(r) its derivatives are calculated R' y R'' and are replaced in (B.1) results

$$r^2 K'' + \left(2r - 2cr^2\right) K' + \left[\left(2Za^{-1} - 2c\right)r - \ell(\ell+1)\right] K = 0$$

(B.4)

Following the regular procedure in the solution of differential equations we must make the radial wave function remain finite throughout the range of variation of the variable r, for this we look for a new substitution of the following form, where s is an integer.

$$\boxed{K(r) = r^s M(r) = r^s \sum_{j=0}^{\infty} b_j r^j = r^s \left[b_0 + b_1 r^1 + b_2 r^2 + \ldots\right]}, \text{ donde } b_0 \neq 0$$

(B.5)

(Frobenius theorem, there is a singular point in $r = 0$ whose first term b_0 is non-zero. In addition, we want the radial function not to cancel at the origin and be continuous case that only applies to the wave s:This is level n = 0!)

We calculate the derivatives K' y K'' and substitute in (B.4)), it turns out

$$r^2 M'' + \left[(2s+2)r - 2cr^2\right] M' + \left[s^2 + s + \left(2Za^{-1} - 2c - 2cl\right)r - l(l+1)\right] M = 0$$

(B.6)

110

How $M(0) = b_0 \neq 0$, $M'(0) = b_1$, y $M''(0) = 2b_2$, for $r = 0$ Is left

$$b_0 \left[s^2 + s - l(l+1) \right] = 0 \quad (l \equiv l) \quad \text{(B.7)}$$

how $b_0 \neq 0$ We are left with the quadratic equation in the unknown s of roots

$$\boxed{s = +l, \quad s = -(l+1)} \quad \text{(B.8)}$$

Note that the interaction of the radial wave of the electron at a distance r from the nucleus up to r =0. These roots are examined according to the physical behavior of the wave function.

IMPORTANT NOTE

Reduced radial function and roots of opposite sign to (B.8)).

Instead of function R(r) in the radial differential equation (B.1) it is convenient to introduce the function $\chi(r)$ (see bibliography 4 and 6: Levich and del Rio)

$$R(r) = \frac{1}{r} \chi(r) \quad \text{(B.9)}$$

Taking the first and second derivatives of R(r) entering in (B.1) we arrive at

$$\frac{d^2 \chi(r)}{dr^2} + \left\{ \frac{2\mu}{\hbar^2} E - \frac{2\mu}{\hbar^2} \left[V(r) + \frac{\hbar^2}{2\mu} \frac{\ell(\ell+1)}{r^2} \right] \right\} \chi(r) = 0$$

(B.10)

What's inside the bracket [ ] is the effective potential energy and equation (A.2) reduces to the equation of the motion of the electron in one dimension. . Not detailed V(r) An interpretation of the behavior of the radial function can be reached near the origin and at great distances. When $r \to 0$ The centrifugal factor $1/r^2$ grows exaggeratedly compared to $E\chi(r)$ y $V(r)E\chi(r)$ with which the differential equation is obtained

$$\frac{d^2\chi(r)}{dr^2} - \frac{\hbar^2}{2\mu}\frac{\ell(\ell+1)}{r^2}\chi(r) = 0 \qquad (B.11)$$

In this differential equation note that it does NOT appear V(r) [R(r) is independent of V(r) if r tends to zero] And only the interaction of the S radial wave of the antielectron (or electron) with the nucleus will occur. The solution of this differential equation is achieved by expressions of the form $\chi(r) = Ar^s$ (See bibliography 4). Substituting this expression into equation (B.11) we obtain the roots

$$\boxed{s = -l, \qquad s = +(l+1)} \qquad (B.12)$$

[Compare with roots (B.8)].

In both cases, physicists reject negative roots (only when they are normalized) because it makes infinity $R(r)$ cuando $r \to 0$.

In our case, as we shall see, they have a periodic and physical meaning.

Therefore, at small distances $\chi(r)$: r^{l+1}

and the radial wave function (B.9) is at short distances from the antinucleus (or nucleus) as

$$R(r) = A r^l \quad \text{(B.13)}$$

Combining (B.13), (B.3) and (B.5) gives us for the complete radial factor

$$R(r) = e^{-cr} r^l \sum_{j=0}^{\infty} b_j r^j \quad \text{(B.14)}$$

Notice, therefore, that the radial function and, as we shall see, the energy will depend on the quantum number of angular momentum, the exponent in (B.13) or (B.14) of $l = 0,1,2,3,\ldots n-1$ and not of the roots as such.

Now we can summarize these results for the 4 roots as follows:

$$s = \begin{cases} +l & \text{si } l = 0, \quad +(l+1) \;\; \forall l = 0,1,2,3,\ldots(n-1) \\ -l & \text{si } l = 0, \quad -(l+1) \;\; \forall l = 0,1,2,3,\ldots(n-1) \end{cases} \quad \text{(B.15)}$$

Aftermath

As we noticed, the s-counter is associated or connected with the angular momentum quantum number. $l \equiv \ell = 0,1,2,3,\ldots,n-1$ which in turn relates to the energy levels of hydrogen atoms. For $l = 0$ (**s**=0) (states s, not to be confused with the s of the quantifier) the wave function of the electron interacts with the nucleus and the nucleus can only be found at a distance r from it because of the centrifugal potential in the equation (B.10); Consequently, there is a space for the S wave between the electron and the nucleus corresponding to a zero energy level (n = 0) as we will see. (It is physically unacceptable for the electron to be in the nucleus. At short distances the calculation corresponds to the electronic wave s as seen by the graph of the radio wave function s. The probability of finding the electron at a distance r from the wave s of spherical symmetry and independent of the angles θ, ϕ is given by $|R|^2 r^2 dr$. From (B.13) we have that for small values of the distance r between nucleus and electron the probability is proportional to $r^{2(l+1)} dr$ and means that the probability near the nucleus of the radio wave is higher and decreases the higher is l = 1, 2, 3,... The consequence is that the centrifugal potential throws the electron away from the nucleus which does not allow it to collapse with it. Finally we find two methods to solve the radial equation and we find something curious but explainable in relation to the roots: while one method gives us negative the root in another is positive and, as will be seen, it is due to two symmetric systems, for matter and antimatter.

-- End of note.

From the preceding note we conclude that the calculation leading us to the energies of the matter and antimatter system and the respective radial functions depends on l = 0, 1, 2, 3,...n − 1. Consequently, with s = l equation (B.6) becomes: (see Article 2, where parallel development is made for the matter and antimatter system)

Replace

$$rM'' + [2\ell+2 - 2cr] M' + [2Za^{-1} - 2c - 2c\ell] M = 0$$

(B.16)

Here we make the following consideration: according to (B.5) and explained for equation (B.18)) $M(0) = b_0 \neq 0$, and

$$c = \sqrt{\frac{-2E}{ae^2}}, \text{ It is fulfilled that:}$$

$$\left[Za^{-1} - c - cl \right] = 0 \qquad \text{(B.17)}$$

This is the ratio that determines the energy levels of hydrogen atoms, including the vacuum interaction s = 0; It will lead to any periodic system according to the sequence chosen. By applying (B.15) to (B.17) we calculate the n≥0 energy levels of matter and antimatter atoms.

Note: in the specialized literature, see the bibliography or consult the one of interest, following the calculation process one arrives at the distribution of the energy levels of the system (see bibliography 4, Eq. 38.15) expressed a little more elaborated as follows:

$$n_r + l + 1 = \frac{Z}{ac} \quad \wedge \quad a = \frac{\hbar^2}{me^2} \quad y \quad c = \sqrt{\frac{-2E_n}{ae^2}} \qquad \text{(B.18)}$$

Here $\hbar, m$ y e are the known constants of atomic physics. We express this relationship in another way given by quantum numbers. n_r, l y n : $n_r + l + 1 = n$ y n_r is the radial quantum number that determines the number of nodes of the radial function. For a value when fixing the quantum number of angular momentum it is fulfilled that $n \geq l + 1$, consequently

$Za^{-1} = c(l+1) \Rightarrow Za^{-1} - c - cl = 0$ and (B.18) is precisely (B.17).

C. Roots obtained (B.15) are substituted in (B.17) for energy levels

Next, we write these corresponding roots with the levels of the atom for the states $l = 0\,(s)$; $1\,(p)$; $2\,(d)$; $3\,(f)$;... which we will match with the periods of the system of hydrogen atoms figure C.1, matter and antimatter, as follows:

$$s \equiv l = \begin{cases} +n \text{ si } n = 0, & +(n-1) \; \forall = n = 1,2,3,...,(l+1): \text{ Materia} \\ -n \text{ si } n = 0, & -(n+1) \; \forall n = -1,-2,-3,...,-(l+1): \text{ Antimateria} \end{cases}$$

(C.1)

It is now possible to determine the energy levels n of the hydrogen atom by substituting the counter s (C.1) in (B.17) for matter and antimatter as shown below:

$$\text{MATERIA }(l = +n = 0)$$
$$|z|a^{-1} - c - cn = 0$$
$$\frac{|z|}{ac} = (1+n)$$

$\leftarrow (\text{onda s interactuando con el núcleo}) \rightarrow$

$$\text{ANTIMATERIA }(l = -n = 0)$$
$$|z|a^{-1} - c + cn = 0$$
$$\frac{|z|}{ac} = (1-n)$$

(C.2)

Note that the quantization condition [Z/a implies that c] $(1+n) = (1-n)$ is satisfied only for the zero-energy level, n = 0 which corresponds to the quantum number $l = 0$ for state s. We square and clear the energy of the constant $c = \pm\sqrt{\dfrac{-2E}{ae^2}}$ results for level zero:

$$E_{n=0} = -\dfrac{|Z|^2 \times 2.18 \times 10^{-18}}{(1+n)^2} \text{ ergios} \qquad\qquad E_{n=0} = -\dfrac{|Z|^2 \times 2.18 \times 10^{-18}}{(1-n)^2} \text{ ergios}$$

(C.3)

Now we substitute in (B.17) the roots $l = +(n-1)$ y $l = -(n+1)$:

MATERIA $\left[l = (n-1)\right]$, $l \geq 0$ y $n \geq 0$ \qquad\qquad ANTIMATERIA $\left[l = -(n+1)\right]$

$|Z|a^{-1} - c - c(n-1) = 0$ \qquad\qquad $|Z|a^{-1} - c + c(n+1) = 0$

$\dfrac{|Z|}{ac} = n$ subcapas $\forall n = 1,2,3,...$ \quad ← subcapas → \quad $\dfrac{|Z|}{ac} = -n$ $\forall n = -1,-2,-3,...$

(C.4)

Note that here n is negative for periods in antimatter and thus positive for atomic levels (see "The New Quantum Mathematics of the Periodic Table" (320 pages) and papers 2, 3 and 4 published on Amazon). In any case, when squared, the n is positive. Is left

MATERIA $\left[l = (n-1) \; \forall n \geq 1 \right]$ ANTIMATERIA $\left[l = -(n+1) \; \forall n \leq -1 \; (\text{periodos}) \right]$

$$E_n = -\frac{|z|^2 \times 2.18 \times 10^{-18}}{n^2} \; ergios \qquad E_n = -\frac{|z|^2 \times 2.18 \times 10^{-18}}{n^2} \; ergios$$

(C.5)

In both cases it is not the root as such with physical meaning but the quantum number of angular momentum that remains equal to $\ell = 0, 1, 2, 3, \ldots, n-1$ for both systems. The denominator n^2 is the number of degenerate eigenfunctions that all correspond to equal energy. E_n for n = 1, 2, 3,… We can dispense with the absolute value since for a positive and / or negative nucleus when squared it gives positive and does not affect the minus sign.

D. New _periodic quantum energy relationships

The expressions for the energy levels of the atomic orbitals (C.3) and (C.5) are summarized in one, both for matter and antimatter as shown below with an example where the energy of the electron is calculated at a level if not, in addition, the orbitals at that level.

matter

$$E_\alpha = -\frac{z^2 \times 13.6 \; eV}{\left\{ (\alpha+1) + \sum_{l=1}^{n-1} (2l+1)[\alpha-l] \right\} - \left\{ \alpha + \sum_{l=1}^{n-1} (2l+1)[\alpha-(l+1)] \right\}}, \; \alpha = 0, 1, 2, 3, \ldots ($$

(D.1)

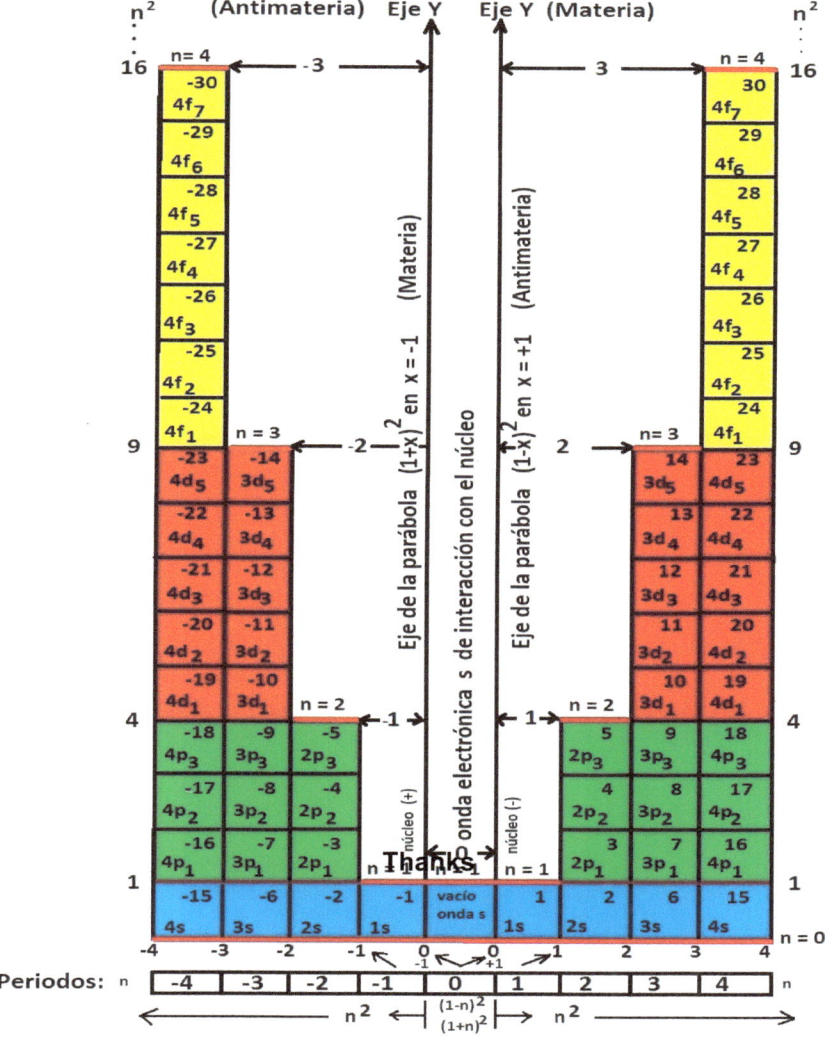

Figure C.1 The roots (2.30) obtained by two ways in the solution of the radial wave equation correspond to the periods, as shown here. The quantum impulse number $l = 0,1,2,3,...,n-1$ In (2.30) it is substituted in the relation (2.29) and we obtain the height of the negative and positive periods, this height are the atomic orbitals and the degenerate energies at each level n (the n of the stepped function). Note that the level n = 0 coincides with the X-axis and see the abscissas -1 and +1 obtained from the denominator of the energy relations for this system the origin 0. Each system has its respective periodic functions. Conclusion: THE SUBJECT IS DESIGNED FROM THIS SYSTEM.

This relationship determines the energy from the zero level to the n=1.2, 3,..., level and the orbitals at each level. The denominator in (D.1) is the function T(α) which determines the number and type of orbitals at each level of hydrogen atoms. Note below in Figure (C.1) and following that the variable alpha, a, We will make it correspond to the periods that in turn correspond to the abscissae, X, and that overlap with the energy levels n: $n = l + 1 \Rightarrow l = n - 1$:

$$T(\alpha) = \left\{ (\alpha+1) + \sum_{l=1}^{n-1}(2l+1)[\alpha-l] \right\} - \left\{ \alpha + \sum_{l=1}^{n-1}(2l+1)[\alpha-(l+1)] \right\}$$

(D.2)

Note that $a = 0$ corresponds to zero angular momentum and zero period (see Figure C.1)

For example, the energy of a Z proton atom for an electron at the n = 3 level and the orbitals that electron can occupy is calculated from (D.1) and is first done with the denominator (D.2):

$$T(3) = \{(3+1) + 3(3-1) + 5(3-2)\} - \{3 + 3(3-2)\} = 15 - 6 = 9 \text{ orbitales para } n = 3$$

$$\begin{Bmatrix} \emptyset s^1 \\ 1s^1 \\ 2s^1 \quad 2p^3 \\ 3s^1 \quad 3p^3 \quad 3d^5 \end{Bmatrix} - \begin{Bmatrix} \emptyset s^1 \\ 1s^1 \\ 2s^1 \quad 2p^3 \end{Bmatrix} = \{3s^1 \quad 3p^3 \quad 3d^5\} = 9 \text{ orbitales} = 3^2$$

Notice how we have obtained the type of orbitals at the n = 3 level. If we go to the periodic system for hydrogen atoms, figure C.1, we count 15 squares from period zero to period three (level n = 3) and count 6 squares from zero to level n = 2 and the difference gives us 9 orbitals that are 3s, 3p y 3d. Now: each square corresponds to an orbital function including the square in the labeled zero period 0s¹ since the S orbitals are the only ones that interact with the nucleus. The electron is at the level n = 1 labeled 1s¹ and the kernel set, wave 0s¹ y 1s¹ they constitute the minimum energy of the atom calculated with (D.1). For the orbital energy at the level n = 3 we now substitute in (D.1) the result T(3), it is

$$E_{n=3} = -\frac{Z^2 \times 13.6 \, eV}{\{(3+1)(1)+3(2)+5(1)\}-\{3(1)+3(1)\}} = -\frac{Z^2 \times 13.6 \, eV}{3^2}$$

We see, then, that these new quantum-periodic relations go beyond the Bohr relation and the results of Schroedinger's quantum mechanics. Now let's calculate for antimatter.

Antimatter (alfha, α, indicates periods that correspond to n levels of the atom)

$$E_\alpha = -\frac{Z^2 \times 13.6 \, eV}{\left\{\alpha + \sum_{l=1}^{n-1}(2l+1)[\alpha+(l+1)]\right\} - \left\{(\alpha-1) + \sum_{l=1}^{n-1}(2l+1)[\alpha+l]\right\}}, \quad \alpha = 0,-1,-2,...$$

(D.3)

Note that the denominator in (D.3) calculates not only the orbitals but the type of orbitals of a hydrogen atom of antimatter. α = -3 (ℓ = 2 y n = 3, according to C.1) we express as a function that gives us the number of eigenor orbital functions at the level n = 3 (ℓ = 2) that overlaps with period -3:

$$\bar{T}(\alpha) = \left\{ \alpha + \sum_{l=1}^{n-1}(2l+1)[\alpha+(l+1)] \right\} - \left\{ (\alpha-1) + \sum_{l=1}^{n-1}(2l+1)[\alpha+l] \right\}$$

(D.4)

how α = -3 we develop (D.4) up to $l = 2$:

$$\bar{T}(\alpha) = \{\alpha + 3(\alpha+2) + 5(\alpha+3)\} - \{(\alpha-1) + 3(\alpha+1) + 5(\alpha+2)\};$$

later

$$\bar{T}(-3) = \{-3 + 3(-3+2) + 5(-3+3)\} - \{-3-1 + 3(-3+1) + 5(-3+2)\}$$

$$\bar{T}(-3) = \{-3 + 3(-1)\} - \{(-4) + 3(-2) + 5(-1)\} = \{-6\} - \{-15\} = 9 \text{ orbitales}$$

$$\begin{bmatrix} 0s^{-1} \\ 1s^{-1} \\ 2s^{-1} \quad 2p^{-3} \end{bmatrix} - \begin{bmatrix} 0s^{-1} \\ 1s^{-1} \\ 2s^{-1} \quad 2p^{-3} \\ 3s^{-1} \quad 3p^{-3} \quad 3d^{-5} \end{bmatrix} = 3s^{+1} \quad 3p^{+3} \quad 3d^{+5}$$

Notice, as for matter, the relationship between the period α = -3 and the energy level of the hydrogen antiatom n = 3. In general, the ratio D(4) reduces to n^2 orbitals with the particularity that the level n = 1 of the stepped function in Figure C.1 for antimatter is constituted by the periods α = 0 y α = -1. Also, notice in the system in Figure C.1 for antimatter that we count 15 frames from zero to period -3 and count 6 frames from zero to period -2 and whose difference gives us the number (positive, of course) of orbitals of a hydrogen antiatom at the energy level n = 3. How to interpret the negative exponent? The result obtained for the matter gave us $3s^1$, $3p^3$ y $3d^5$ and the exponents are positive which corresponds to the negative electrons; In the case of antimatter the exponents are negative that is only explained if the positrons are positive which the reading is that at the level n = 3 of a hydrogen antiatom you have an S orbital (positron can be in this orbital); 3 p orbitals and 5 d orbitals where the positron (negative exponent) can be placed with the energy of level n = 3 calculated with the ratio (D.4). Note that the systems represented in Figure C.1 refer to orbitals and their parameter is R = 1 which is the length of each level n of the step function. The periodic table, R = 2, results from doubling the orbitals of the hydrogen system, R = 1, and placing R = 2 electrons in each orbital.

E. How the periodic hydrogen and periodic table systems of the denominator of energy relations are constructed

To design the hydrogen system, the denominator of energy relationships is based on. (C.3) y (C.5): (note: $\alpha \equiv x = periodos$)

Antimatter:

$(1-n)^2$ sin $\equiv \alpha = 0$ y n^2, si $n \geq 1$, periodos $\alpha = -1, -2, -3, ...$

Matter: $(n+1)^2$ sin $= 0$ $(\alpha = 0)$ y n^2, $n \geq 1$, $\alpha = 1, 2, 3, ...$

Let us match the n-levels of the atom by the abscissae. To this end, it must be fulfilled, first, in the case of the antimatter system, that

$(1-x)^2 = n^2 \Rightarrow (1-x) = n \Rightarrow \boxed{x = 1-n}, \quad \forall n \geq 0$ \hfill (E.1)

Then we have

$x^2 = n^2 \Rightarrow \sqrt{x^2} = \sqrt{n^2} \Rightarrow -x = n \Rightarrow \boxed{x = -n} \quad n \geq 0$ \hfill (E.2)

Note that we are analyzing the antimatter system where abscissas are negative but always the n levels are positive, as it should be.

Hence the unit intervals, I, for the antimatter system as the sublevels of each level are built, always starting with the orbitals. ns, $\forall n \geq 0$ (see figure E.1):

$$I\left[(1-n)-n\right], \forall n \geq 0 \qquad \text{(E.3)}$$

Note that this is a quantum interval of length R = 1. In addition, it is measured contrary to conventional nomenclature and there is no value between the ends of each interval. Some intervals are:

n	$I\left[(1-n),-n\right]$	n	$I\left[(1-n),-n\right]$	n	$I\left[(1-n),-n\right]$
0	$I(1,0)$	1	$I(0,-1)$	2	$I(-1,-2)$
3	$I(-2,-3)$	4	$I(-3,-4)$	5	$I(-4,-5)$, etc.

Note the extremes of each interval (see Figure E.1).

Matter system

As the previous case let us correspond the orbital functions n^2 of each level of the hydrogen atom with the abscissas on the X axis; It is first true that

$$(1+x)^2 = n^2 \implies (1+x) = n \implies \boxed{x = n-1}, \forall n \geq 0 \qquad \text{(E.4)}$$

Then we have

$$x^2 = n^2 \implies \sqrt{x^2} = \sqrt{n^2} \implies +x = n \implies \boxed{x = n} \quad n \geq 0 \qquad \text{(E.5)}$$

We obtain the unit intervals, I, for the matter system as the sublevels of each level are built, always starting with the orbitals. ns, $\forall n \geq 0$ (see Figure E.1):

$$\iota[(n-1)n], \forall n \geq 0 \quad \text{(E.6)}$$

Note the ends of each interval. The length of each interval is, like antimatter, R = 1, and there is no value between the ends of each of the intervals. Some discrete or quantum intervals are:

n	$\iota[(n-1),n]$	n	$\iota[(n-1),n]$	n	$\iota[(n-1),n]$
0	$\iota(-1,0)$	1	$\iota(0,1)$	2	$\iota(1,2)$
3	$\iota(2,3)$	4	$\iota(3,4)$	5	$\iota(4,5)$, etc.

Note the extremes of each interval (see Figure E.1).

Antimatter System

Right ends:

$$[(1-n)n^2], \forall n \geq 0 \equiv [x,(1-x)^2], \forall x = 1-n \quad \text{(E.7)}$$

Far left: (critical coordinate: (1,1), see Figure E.1)

$$[-n,n^2], \forall n \geq 0 \equiv [x,x^2], \forall x \leq 0 \Rightarrow x = -n \quad \text{(E.8)}$$

Matter system

Far left: (critical coordinate: (-1,1), see Figure E.1)

$$[(n-1)n^2], \forall n \geq 0 \equiv [x,(x+1)^2], \forall x = n-1 \quad \text{(E.9)}$$

Right ends: (Figure E.1)

$$[n,n^2], \forall n \geq 0 \equiv [x,x^2], \forall x \geq 0 \Rightarrow x = n \quad \text{(E.10)}$$

Note, in Figure E.1, the construction of the intervals of each system is determined by the denominators of the energy relations and that they only touch the extremes of these intervals and are quantum intervals in the sense that there is no value between the extremes. Here's a brief calculation for the matter system:

Quantum scale

$$T(n) = \begin{cases} (n+1)^2, & n = 0 \\ n^2, & \forall n \geq 1 \end{cases} \quad \text{intervalo:} \quad [(n-1), n] \; \forall n \geq 0$$

This gives us the eigenor orbital functions of atoms at each energy level. n≥0:

$$T(0) = \{(0+1)^2 = 1 \text{ ondas de interacción con el núcleo}\} \text{ intervalo: } |[(0-1),0] = /[-1,0]$$

$$T(1) = \{1^2 = 1 \text{ orbital s}\} \quad \text{intervalo: } |[(1-1),1] = /[0,1]$$

$$T(2) = \{2^2 = 4 \text{ orbitales, 2s y 3 orbitales p}\} \; /[1,2], \text{ etc.}$$

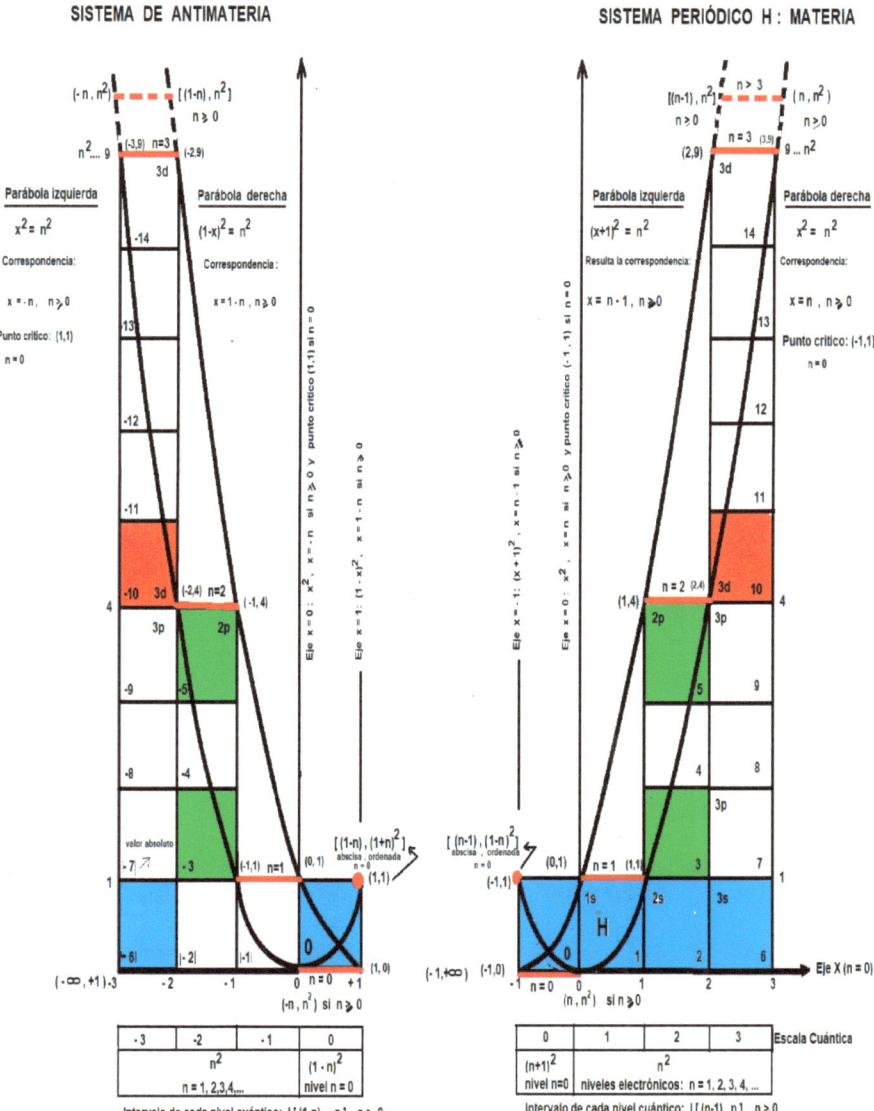

Figure E.1 Quantum-periodic geometry derived from energy ratios deduced from the wave equation leading to a new relationship determining energy levels is illustrated here. Note the abscissa at the critical points: $(1-n)$ y $(1+n)$ si $n=0$ and ordered Y = 1 in both systems.

Cartesian scale

Note the discrete quadratic function for each parabola with the particularity of including the vertex of the parabola $(x+1)^2$ y x^2. Note that we use the variable x that corresponds to the abscissas but that the coordinates (x, y) are a function of the variable n (energy levels). We have for each discrete parable: $x+1)^2$

$$T(x) = \{(x+1)^2\} \text{ de coordenadas } A_n\left[(n-1), n^2\right], \forall n \geq 0 \Rightarrow x = n-1; \quad y = n^2$$

$T(-1) = 0, \Rightarrow n = 0$ donde $A_0[-1, 0]$ (vértice)
$T(0) = 1, \Rightarrow n = 1$ y $A_1[0, 1]$ (altura del nivel n = 0)
$T(1) = 4 \Rightarrow n = 2$ y $A_2[1, 4]$ (altura del nivel n = 2) ; etc.

For the discrete parable x^2 we have

$$T(x) = x^2 \quad \text{de coordenadas} \quad B_n\left[n, n^2\right], \quad \forall n \geq 0 \Rightarrow x = n; \; y = n^2$$

$T(0) = 0^2 \Rightarrow n = 0$ donde $B_0[0, 0]$ (vértice)
$T(1) = 1^2 = 1 \Rightarrow n = 1$ y $B_1[1, 1]$ (un orbital en el nivel n = 1)
$T(2) = 2^2 = 4 \Rightarrow n = 2$ y $B_2[2, 4]$ (4 orbitales en el nivel n = 2) ; etc.

Note that the coordinates corresponding to the two discrete parabolas are the extremes of the levels of the stepped function whose height indicates the proper or orbital functions available in the atoms. Note also the vertices V(-1,0) of the parabola (x+1)² y V(0,0) of x² fix the energy level n = 0 and the curves intersect at I(-1,0) and whose

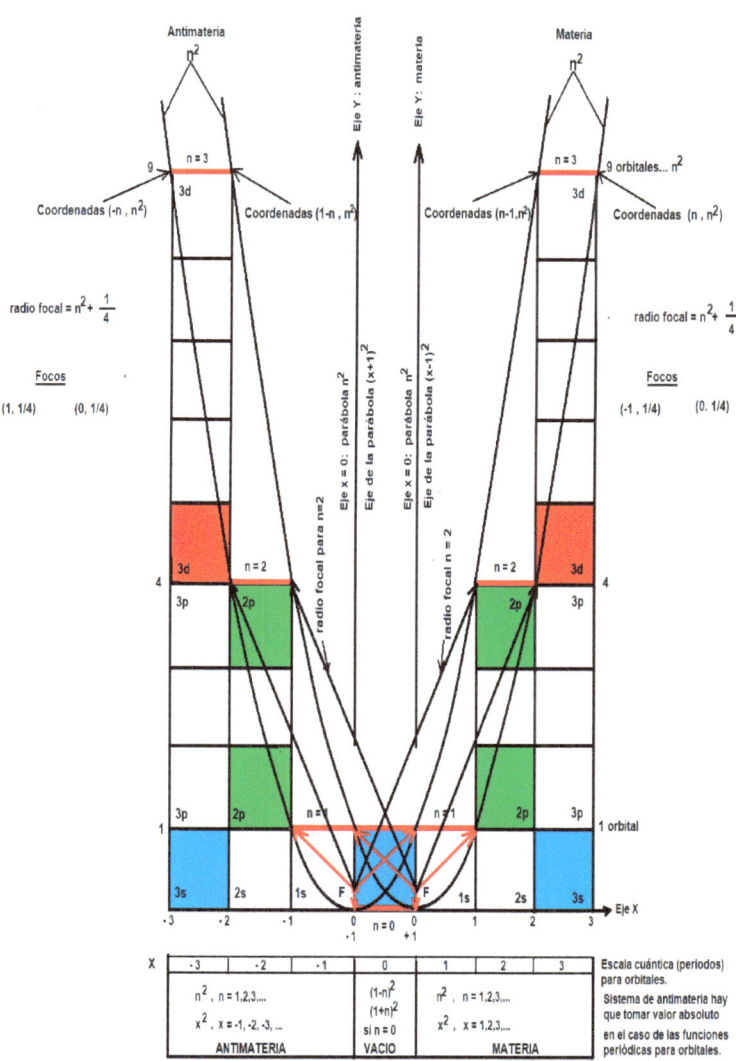

Figure E.2 Here the two systems that are separated in Figure D.1 have been joined. Note the square corresponding to the zero energy level in the interval I(0,0) or the equivalent I(-1,1) There we see the four focal radius that build the unit area square corresponding to the s-orbital of the electron interacting with the nucleus through the vacuum... Each square of unit area represents one orbital. The focal radius is unique to each system and depends on the parameter of each quantum parabola that corresponds to the denominator of the energy function.

coordinates correspond to the points that set the height of the zero level where the four coordinate points (see Figure E.2) make a square of unit area corresponding to an entity or proper or orbital function of the hydrogen atom. It is of interest, and in fact is what characterizes the parabolas, their straight side 4P where each periodic system is unique by its parameter P and no two systems are the same. The vertices V(-1,0) of the parabola $(x+1)^2$ and V(0,0) of x^2 set the energy level n = 0 and focus, respectively, F(-1, ¼) y F(0,1/4). The first directed to the far left of each quantum coordinate level [(n-1) , n^2] and the second of coordinates [n , n^2] directed to the far right as illustrated in figures E.2 and E.3 (see relationship E.14).

Periodic Table Design

To move to the periodic table, the levels of the hydrogen system are doubled, Figure D.1. The level n = 1 of the hydrogen system is duplicated and from the level n≥2 These are duplicated as shown in Figure 1.2 resulting in the periodic table. In general, the levels of the hydrogen system can be doubled, tripled, quadrupled, etc. up to the periods

$$\partial_{n+1} = \begin{cases} 0 & \text{si } n = 0 \\ Rn - 1 & \forall n \geq 1 \quad \wedge \quad R \geq 2 \end{cases} \quad \text{(E.11)}$$

Resulting in the following lower limits of integration:

$\partial_1 = 0; \quad \partial_2 = R - 1; \quad \partial_3 = 2R - 1; \quad \partial_4 = 3R - 1; \quad \partial_5 = 4R - 1;....$ etc. **(E.12)**

The heights or orbitals of the designed system, R = 2 for the periodic table, are calculated with the ratio (1.27), (see section 8.1). Then we match the periods with the abscissas and the levels for the initial periods, x0, and the end, x, as follows:

$$T(x_0) = \frac{(x_0+2)^2}{2^2} = n^2 \Rightarrow \alpha_0 \equiv x_0 = 2(n-1) \quad \text{si } n \geq 1 \quad \textbf{(E.12)}$$

$$T(x_f) = \frac{(x_f+1)^2}{2^2} = n^2 \Rightarrow \alpha \equiv x_f = 2n-1 \quad \text{Si } n \geq 1 \quad \textbf{(E.13)}$$

The discrete or quantum intervals for the periodic table are determined from (E.12) and (E.13), it turns out

$$I[(2n-3),(2n-1)] \forall n \geq 1 \quad \textbf{(E.14)}$$

The reader can draw the above two parabolas on Figure 1.2 and check the quantum or discrete coordinates corresponding to the ends of each level of the step function:

$$[2(n-1), 2n^2], \forall x_0 \equiv \alpha_0 = 2(n-1), n \geq 0 \quad \textbf{(E.15)}$$

Even periods, including vertex (-2,0) if n=0.

For odd periods we have the coordinates

$$[2n-1, \ 2n^2], \forall x \equiv \alpha = 2n-1, n \geq 0 \quad \textbf{(E.16)}$$

By simple geometry the focal radius or radius vector pointing from the focus, F, is calculated for the even states to the coordinates.

$$F(-2, \frac{1}{2}) \quad \text{a las coordenadas} \quad [2(n-1), 2n^2], \quad \forall n = 1,2,3,4,\ldots$$

$$\textbf{(E.17)}$$

Applying the distance between two points results in.

$$\text{radio focal} = r_f = 2\left[n^2 + \frac{1}{4}\right] = 2n^2 + \frac{1}{2} \qquad \text{(E.18)}$$

The expression inside the bracket is the radius vector of the hydrogen system and (1/4) is the parameter of this system which implies that the parameter for the Periodic Table is double, i.e. P = 2(1/4) = ½ and, therefore, the radius vector of the Periodic Table system is twice that of the hydrogen system. In general, for systems R≥2 the radius vector is R times the radius vector of the hydrogen system and the parameter R(1/4).

Final periods of the periodic table

Like even periods, the radius vector is calculated:

$R\left(-1, \frac{1}{2}\right)$ a las coordenadas $\left[(2n-1), 2n^2\right]$, $\forall n = 1, 2, 3, 4, \ldots$ (E.19)

result

$$\text{radio focal} = r_f = 2\left[n^2 + \frac{1}{4}\right] = 2n^2 + \frac{1}{2} \qquad \text{(E.20)}$$

And, the two radius vectors are directed at the ends of each quantum level, Figure 1.2.

Note that the number of periods R under the n of the step function depends on the sequence selected and that it is the straight side of the discrete parabolas: R = 4P; in summary: the quantum number of the periodic table is 1/2 that is given by the parameter P = 1/2 that gives the opening or sequence of the system and that adjusts the system so that the periodic length is repeated according to the largest quantum number of angular momentum. **l = n-1 and allows you to adjust the even and odd periods given by the relation (E.15) and (E.16).**

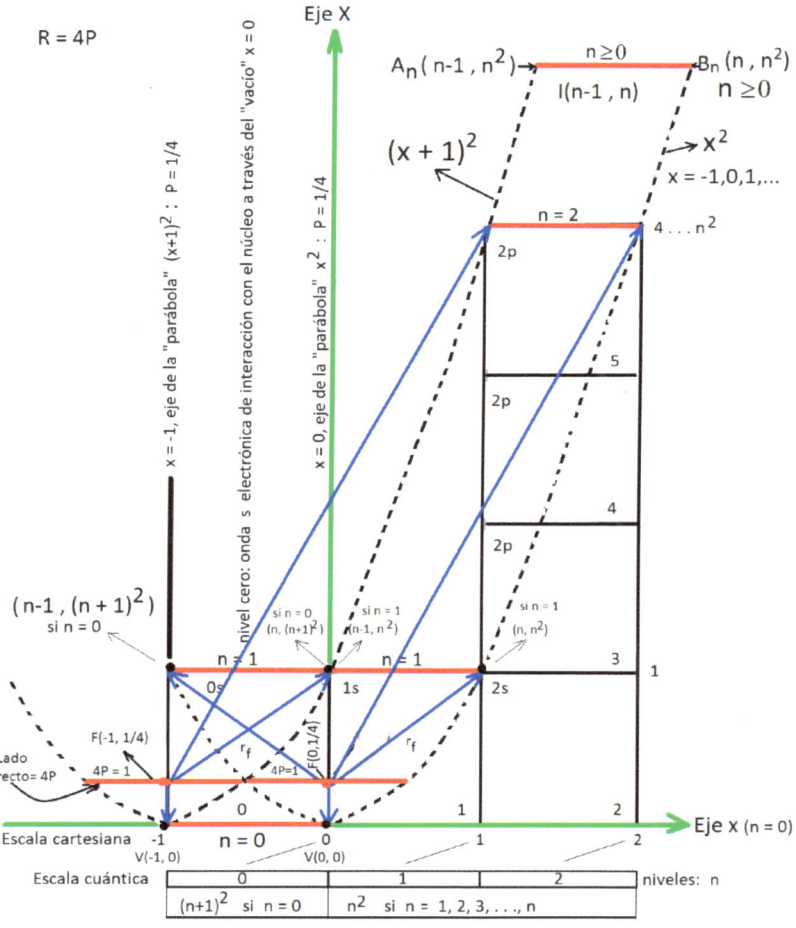

SISTEMA PERIÓDICO DE ÁTOMOS HIDROGENOIDES

Figura E.3 Nótese, en la escala cuántica, el periodo o nivel cero se construye con el radio focal dirigido a los vértices del cuadrado de área unidad, el orbital $0s^1$, luego cada radio vector discreto va construyendo los n^2 orbitales a partir del nivel cero. Además, se ha dibujado en línea punteada las parábolas discretas que sirven como una especie de "ruta" de las ondas electrónicas por donde se desplaza el lado recto y los círculos osculatrices.

F. Calculation of the general term of the succession formed by the number of orbitals of the hydrogen system in the design of the Periodic Table

How many orbitals should be taken from the hydrogen system and in what order to design the Periodic Table or other system? What quantum-periodic algebra tool is required for its calculation?

Periodic systems are cyclic, Figure E.1, where each period begins with an s orbital and ends in p orbitals for left-s-stepped systems such as the periodic table. The orbital functions of the hydrogen system on the left are as follows (see 320-page text):

$$s(n) = n + \sum_{l=1}^{n-1}(2l+1)[n-(l+1)] \qquad \forall n \geq 0$$

$$\overset{\leftarrow}{p}(n) = \sum_{l=0}^{n-1}(2l+1)[n-l] \qquad \forall n \geq 2$$

$$\overset{\leftarrow}{d}(n) = n + \sum_{l=1}^{1}(2l+1)[n-(l+1)] + \sum_{l=2}^{n-1}(2l+1)[n-l] \qquad \forall n \geq 3$$

$$\overset{\leftarrow}{f}(n) = n + \sum_{l=1}^{2}(2l+1)[n-(l+1)] + \sum_{l=3}^{n-1}(2l+1)[n-l] \qquad \forall n \geq 4$$

$$\overset{\leftarrow}{g}(n) = n + \sum_{l=1}^{3}(2l+1)[n-(l+1)] + \sum_{l=4}^{n-1}(2l+1)[n-l] \qquad \forall n \geq 5$$

$$\overset{\leftarrow}{h}(n) = n + \sum_{l=1}^{4}(2l+1)[n-(l+1)] + \sum_{l=5}^{n-1}(2l+1)[n-l],... \qquad \forall n \geq 6$$

(F.1)

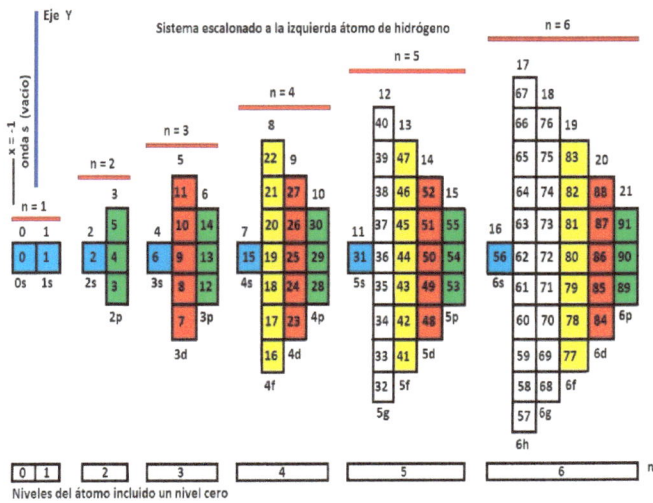

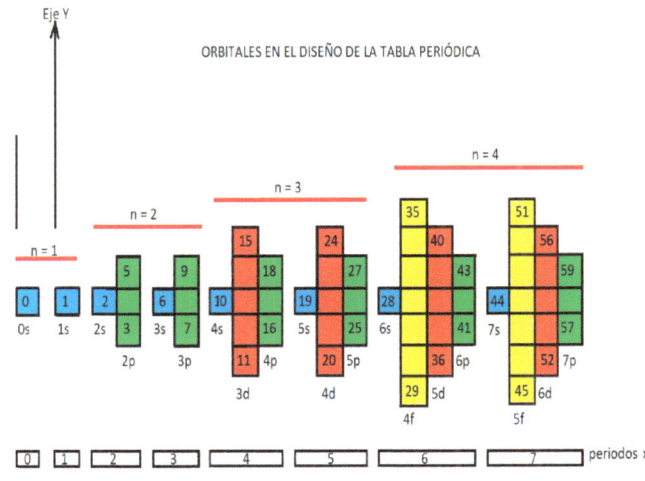

Función orbital = p(x) = x + 3(x-1) + 5(x-3) + 7(x-5), x corresponde con el periodo: x = 0, 1, 2, 3, 4, 5, 6 y 7   (hasta n = 4)

Figura F.1 Arriba el sistema periódico hidrogenoide y abajo el sistema que conduce a la tabla periódica al asignar dos electrones por orbital. Véase el cálculo de cómo se construye la tabla periódica (abajo) del sistema hidrogenoide (arriba)

From (F.1) we go to the generating function of the hydrogen system that corresponds to the p orbitals that closes the cycle at each level n = 0, 1, 2, 3, ... and develop the summation results

$$p(n) \equiv f(n) = 1(n-0) + 3(n-1) + 5(n-2) + 7(n-3) + 9(n-4) + 11(n-5) + \ldots$$

$$p(0) = 1(0-0) = 0$$

$$p(1) = 1(1-0) = 1 \Rightarrow 1s^1$$

$$p(2) = 1(2-0) + 3(2-1) = 5 \text{ orbitales } (1s^1 \ 2s^1 \ 2p^3)$$

$$p(3) = 1(3-0) + 3(3-1) + 5(3-2) = 14 \text{ orbitales } (1s^1 \ 2s^1 \ 2p^3 \ 3s^1 \ 3p^3 \ 3d^5)$$

And we follow the infinite forming the succession

$$0, 1, 5, 14, 30, 55, 91, \ldots \quad \forall \ p(n) \equiv f(n) \ \wedge \ n = 0,1,2,3,4,5,6,\ldots$$

(F.2)

Note that p(0) = 0 since the p orbitals do not interact with the nucleus and n = 0 corresponds to the vacuum; The S wave if it interacts with the nucleus and is written as 0s¹ where zero implies n = 0 and 1, in the superscript s^1, The S orbital

Can a general term of this sequence be calculated? The answer is affirmative and general for infinite periodic systems, how? Up to constant difference of the last row. Explain: Let be the succession

$b_1, b_2, b_3, b_4, b_5, b_6, \ldots$ that we call the main row and we calculate the difference, symbol Δ, from b_2 between each term and the previous one and we obtain the second row and we repeat the process until the last row is equal to the differences, the result is expressed in (F.3). In the specialized literature[9] it is shown that so many combinatorial numbers of numerator (n – 1) and denominators 0, 1, 2, ..., appear or are formed. k rows, multiplied the combinatorial numbers by the first number in each row.

$$
\begin{array}{cccccl}
b_1 & b_2 & b_3 & b_4 & b_5\ldots & = \textbf{fila principal} \\
V_{1,1} & V_{1,2} & V_{1,3} & V_{1,4} & \ldots & = 1^a \text{ fila }, k = 1 \\
V_{2,1} & V_{2,2} & V_{2,3} & \ldots & & = 2^a \text{ fila }, k = 2 \\
& g & g & & & \\
& g & g & & & \\
& g & g & & & \\
V_{k,1} & = V_{k,2} & = V_{k,3} & = \ldots & & = k \text{ - ésima fila}
\end{array}
$$

(F.3)

This gives the general expression (E.4) of the sequence $b_1, b_2, b_3, b_4, \ldots$

$$b_n = b_1 + \sum_{k=1}^{k} \Delta_{k,1}\binom{n-1}{k} = b_1 + \Delta_{1,1}\binom{n-1}{1} + \Delta_{2,1}\binom{n-1}{2} + \ldots + \Delta_{k,1}\binom{n-1}{k}$$

(F.4)

Euler's factorial formula is

$$C_{m,n} = \binom{m}{n} = \frac{m!}{(m-n)!\, n!} \qquad \textbf{(F.5)}$$

Next, we have some combinatorial numbers multiplied by Δ:

$$\Delta_{1,1}\binom{n-1}{1} = \Delta_{1,1}\frac{(n-1)!}{(n-2)!\,1!} = \Delta_{1,1}(n-1)$$

$$\Delta_{2,1}\binom{n-1}{2} = \Delta_{2,1}\frac{(n-1)!}{(n-3)!\,2!} = \Delta_{2,1}\frac{(n-1)(n-2)}{2}$$

$$\Delta_{3,1}\binom{n-1}{3} = \Delta_{3,1}\frac{(n-1)!}{(n-4)!\,3!} = \Delta_{3,1}\frac{(n-1)(n-2)(n-3)}{6}$$

(F.6)

With this information we proceed to calculate the general term of the succession (E.2):

```
0        1        5        14        30        55 ... fila principal
    1        4        9        16        25 ...    1° fila
        3        5        7        9 ...        2° fila
            2        2        2 ...    última fila: diferencias iguales
```

We take the first terms of each row:

$a_1 = 0,\qquad \Delta_{1,1} = 1,\qquad \Delta_{2,1} = 3,\qquad \Delta_{k,1} = 2,\quad k = 3 \Rightarrow$ diferencias iguales

We substitute in (E.4) and calculate with (E.6), it results when making the product:

$$b_n = 0 + (n-1) + \frac{3(n^2 - 3n + 2)}{2} + \frac{2(n^3 - 6n^2 + 11n - 6)}{6}$$

After grouping and canceling common terms, it finally turns out

$$b_n = \frac{n(n-1)(2n-1)}{6}, \quad \forall n \geq 1 \qquad \text{(F.7)}$$

Note that the first term is for n = 1, i.e. $b_1 = 0$ and corresponds to n = 0 in the generating function of the hydrogen system; The second term is , $b_2 = 1$ and corresponds to n = 1 in the generating function; Consequently, it is fulfilled for $f(n), n \geq 0$ y $b_n, n \geq 1$:

$$(n \geq 0): f(n) = (n-0) + 3(n-1) + 5(n-2) + 7(n-3) + \ldots = b_n = \frac{n(n-1)(2n-1)}{6}, \quad \forall n \geq 1$$

(F.8)

Calculation of the general term of sequence 1, 5, 14, 30, 55, ...

We proceed as the sequence 0, 1, 5, 14, 30, 55, ... and you get

$$b_n = \frac{n(n+1)(2n+1)}{6}, \quad \forall n \geq 1 \qquad \text{(F.9)}$$

Now notice that despite obtaining the general term (F.9)) of the sequence 1, 5, 15, 30, ... we can start it from the zero-energy level and (F.8) is:

$$(n \geq 0): f(n) = (n-0) + 3(n-1) + 5(n-2) + 7(n-3) + \ldots = b_n = \frac{n(n+1)(2n+1)}{6}, \quad \forall n \geq 0$$

(F.10)

But now it corresponds to the sequence (F.2) despite obtaining it from the sequence 1, 5, 14, 30, 55, 91, ... The reason is that, as will be seen, for the periodic table, R = 2, or systems in general, $R \geq 2$, one corresponds to the initial periods and the other to the final periods under the n of the step function (see Figure 1.2). Below are the overall results for states s and p $R \geq 2$. For the Periodic Table R = 2:

Initial periods systems in general orbitals s

$$\alpha_0 = R(n-1), \quad \forall n = 1,2,3,4,\ldots, (n \geq 1), R \geq 2$$

$$b_n = \left[R\left(\frac{n(n-1)(2n-1)}{6}\right) \right], \quad \forall n \geq 1, x_0 = R(n-1)$$

orbitales H: 0,1,5,14,30,...

(F.11)

Final periods systems in general orbitals s

$$x \equiv \alpha_f = Rn - 1, \quad \forall n = 1,2,3,4,\ldots, (n \geq 1), R \geq 2$$

$$b_n = \left[R\left(\frac{n(n+1)(2n+1)}{6}\right) - n^2 \right], \quad \forall n \geq 1, x = 2n-1, R \geq 2$$

Orbitales H: 1,5,14,30,...

(F.12)

Next, we rewrite the relationships for the calculation of any periodic system. $R \geq 2$ from the periodic hydrogen system, $R = 1$, for the filling of the orbitals p. once the filling sequence is established, which depends on R, each orbital is divided into $\phi_1 \geq 2$ parts.

Initial periods systems in general orbitals p

$$\alpha_0 = R(n-1), \quad \forall n = 1,2,3,4,\ldots, (n \geq 1), R \geq 2$$

$$p_3(\alpha_0) \equiv b_n = \left\{ \left[R\left(\frac{n(n-1)(2n-1)}{6}\right) + n^2 \right] - 1 \right\}, \quad \forall n \geq 1, \quad \phi_1 \geq 2, \quad R \geq 2$$

(F.13)

Final periods systems in general orbitals p

$$x \equiv \alpha_f = Rn - 1, \quad \forall n = 1, 2, 3, 4, \ldots, (n \geq 1), R \geq 2$$

$$p_3(\alpha) \equiv b_n = \left[R\left(\frac{n(n+1)(2n+1)}{6}\right) - 1 \right], \quad \forall n \geq 1, \quad R \geq 2$$

(F.14)

Calculation of the general term of sequence 1, 1, 2, 6, 15, 31, ... s-orbitals of the hydrogen system: Figure F.1

So far we have encoded the zero box of the zero period and we have shown that it applies to the wave s of the electron in level one and that it interacts with the nucleus between the space of this and the electron that corresponds, precisely, with the zero level. Now we are going to form the sequence where the zero grid corresponds to an orbital function that is the same as level one and we apply (F.4), the general term results:

n	0	1	2	3	4	5...
$a_1 =$	1	1	2	6	15	31...
$V_{1,1} =$	0	1	4	9	16...	$n^2, \forall n \geq 1$
$\Delta_{2,1} =$		1	3	5	7...	$(2l+1), \forall l \geq 0$
$\Delta_{3,1} =$			2	2	2...	$[2(l+1)+1]-[2l+1]=2$

$$b_n = \left[\underbrace{\left(\frac{n(n-1)(2n-1)}{6}\right)+1}_{\text{orbitales s H}:1,1,6,15,31,...} \right], \quad \forall n \geq 0$$

(F.15)

Aftermath

When applying the roots $+l = n, (n = 0)$ y $+l = (n-1)$ in the relation that determines the levels of the atom, (B.17), we obtained for matter the relations of energy (C.3) and (C.5) whose denominator is, respectively,

$(n+1)^2, \forall n = 0$ y $n^2, \forall n \geq 1$. This results in the sequence for the orbitals at levels 0, 1, 2, 3, 4, ..., respectively, 1, 1, 4, 9, 16,... Let's use (F.15) in a way that allows us to obtain this sequence, that is, the orbital functions at each level, what quantum-periodic calculation instrument do we have at hand? The relationship (see Text 320 pages and articles 2, 3 and 4)

$$T(n) = \left\{(n+1) + \sum_{l=1}^{n-1}(2l+1)[n-l]\right\} - \left\{n + \sum_{l=1}^{n-1}(2l+1)[n-(l+1)]\right\}$$

which in abbreviated form is $T(n) = s(n+1) - s(n)$. In (F.15) we change to (n + 1) to avoid double cancellation and change to s(n +1) and the expression s(n),

$$s(n) = n + \sum_{l=1}^{n-1}(2l+1)[n-(l+1)] = n + 3(n-2) + 5(n-3) + 7(n-4) + \ldots$$

It remains unchanged with what results in the relationship that determines the number of orbitals at each energy level, including the zero and orbital level. 0s[1]:

$$T(n) = \left\{\left(\frac{n(n+1)(2n+1)}{6}\right) + 1\right\} - \left\{n + \sum_{l=1}^{n-1}(2l+1)[n-(l+1)]\right\}, \forall n \geq 0$$

$$T(n) = \left\{\left(\frac{n(n+1)(2n+1)}{6}\right) + 1\right\} - \left\{n + 3(n-2) + 5(n-3) + 7(n-4) + \ldots\right\}$$

(F.16)

we have:

$$T(0) = \{(0)+1\} - \{0\} = 0s^1$$
$$\{s^1\} \{0\}$$

(see Figure F.1)

$$T(1) = \{(1)+1\} - \{1\} = 1s^1$$
$$\{s^1 + 0s^1\} \{s^1\} = 1s^1$$

We explain: for T(0) corresponds to the zero level, n = 0, and the wave s interacting with the labeled nucleus 0s¹.

Then, we have the level n = 1, T(1), formed by levels 0 and 1 to which we subtract the level 0s1 and the orbital results 1s¹ to accommodate the electron that is rippling around the nucleus and interacting with it through zero vacuum.

We now move to the energy level n = 2 and obtain:

$$T(2) = \left\{ \left(\frac{2(2+1)(2(2)+1)}{6} \right) + 1 \right\} - \{2\}$$

$$T(2) = \{(5)+1\} - \{2\} = 4 \text{ orbitales: } 2s^1 \ 2p^3 \quad \text{etc.}$$

$$T(2) = \left\{ \begin{array}{l} 0s^1 \\ 1s^1 \\ 2s^1 \ 2p^3 \end{array} \right\} - \left\{ \begin{array}{l} 0s^1 \\ 1s^1 \end{array} \right\} = 2s^1 \ 2p^3$$

As explained, each square corresponds to an orbital. The first square is represented as 0s¹,. 0 is the square code and exponent 1 represents the s-orbital interacting with the nucleus through zero; that is, the space between the nucleus and level one where the electron is located. Note that the zero square of period zero is counted and subtracted to get the number of orbitals, n^2, and the type at each level n. Without the orbital 0s¹ There would be no results that harmonized with the quantum calculations of atoms, conclusion?

Generating function of the Periodic Table

The generating function of the Periodic Table corresponds to the final or odd periods 1, 3, 5, 7, ... given by the relationship $x \equiv \alpha = 2n - 1, \forall n \geq 1$ for being a staggered system to the left. The function for the elements that close each period is given by

$$p_6(x) = 2x + 6(x-1) + 10(x-3) + 14(x-5) + ...$$

$$p_6(1) = 2(He), \quad p_6(3) = 18(Ar), \quad p_6(5) = 54(Xe), \quad p_6(7) = 118, \quad p_6(9) = 218,...$$

We get the sequence 2, 18, 54, 118, 218, ..., Now we apply differences and find constant difference 8. In general, each system is characterized by the constant difference 4R, where R is the number of periods where the periodic length is repeated. So we have:

```
    2        18        54        118        218,...
        16        36        64        100...
            20        28        36...
                 8         8...
```

(F.17)

Applying (F.4), it results

$$p_6(x) \equiv b_n = 2\left[2\left(\frac{n(n+1)(2n+1)}{6}\right) - 1\right], \quad \forall n \geq 1 \quad \text{(F.18)}$$

Note how levels 1, 3, 5 and 7 are designed (although applied to infinite levels) of the elements of the periodic table from the hydrogen system as illustrated in Figure F.1. In the general term (F.18) we find the term of the hydrogen system (F.10) that corresponds to the sequence 1, 5, 14, 30, 55, 91, ..., which begins at level one.

Aftermath

Notice in Figure F.1 how the orbitals of the periodic table are constructed from the hydrogen system. For example, how do you construct the elements He and Ar that complete periods 1 and 3 of the Periodic Table?

answer

Let's look at the relationship between the previous generating function: we have

$x = 2n-1$, if $n = 1$ corresponds to the period $x = 1$, consequently

$$p_6(1) = 2(1) = 2(He) \equiv b_1 = 2\left[2\left(\frac{1(1+1)(2+1)}{6}\right) - 1\right] = 2[2(1) - 1] = 2(He)$$

$$1s^2 \qquad\qquad 2\left[(0s^1 + 1s^1) - 0s^1\right] = 1s^2$$

The level $n = 1$ of the hydrogen system was doubled, which is already duplicated, as shown in figure F.1 and we subtract the square zero, $0s1$, and we have the orbital 1s to which we fill with two electrons and results in He.

If $n = 2$ we have the period $2 \times 2 - 1 = 3$, it is

$$p_6(3) = 2(3) + 6(2) = 18(Ar) \equiv b_2 = 2\left[2\left(\frac{2(2+1)(4+1)}{6}\right) - 1\right] = 2[2(5) - 1] = 18(Ar)$$

$$1s^2\, 2s^2\, 2p^6\, 3s^2\, 3p^6 \;=\; 2\left[\left(0s^1\, 1s^1\, 2s^1\, 3s^1\, 2p^3\, 3p^3\right) - 0s^1\right] = 1s^2\, 2s^2\, 2p^6\, 3s^2\, 3p^6$$

On the left side the generating function gives us the atomic number and configuration of element 18, Ar. On the right side we take 5 orbitals of the hydrogen system that complete the levels n = 1 and 2 of the hydrogen system, see figure F.1, now we double the 5 orbitals and give us 10 counts from square 0 to square 9 of level 3 and subtract the zero square, there are 9 orbitals left to fill with two electrons for a total of 18 electrons to neutralize the nucleus with 18 protons of the element Ar.

Final Note

This first article, together with the new periodic table, is designed as a manual for the teacher of Chemistry, Physics, Biology and Engineering at the middle and university level. Here I have developed the specific calculations for the Periodic Table, but now new formulations have been included and how the Periodic Table has been designed based on the hydrogen system. The most exhaustive and calculative development is found in the three articles published in amazon.com. Finally, the text that summarizes the 4 articles is available, entitled: "THE NEW QUANTUM MATHEMATICS OF THE PERIODIC TABLE" also published in Amazon.com.

There is still much work to be done: to develop the mirror image of matter. There are 32 antimatter figures and very laborious writing functions. Then move on to the new research project on the Chemistry of Life and the Environment (burning fossil fuels).

Thanks

I wish to thank the 50th General Assembly & 47th IUPAC World Chemistry Congress held from July 5-12, 2019-Paris, France for the certificate of communication and attendance. Also, a big thanks to Dr. Jan Reedijk, University of Leiden Netherlands, for his admirable interest in reading the manuscript, article 1, and his helpful advice for its publication. In addition, to the American Chemical Society, ACS, and especially to its President Dr. Luis Echegoyen of UTEP for his suggestions and to the Peruvian Chemical Society for receiving the manuscript and other Chemical Societies in the world to which the work was sent. Finally, to my children Sashah Elíh, Otniel and Nathan for their support. In general, to the WORLD SCIENTIFIC COMMUNITY to which the new quantum-periodic knowledge has been sent.

Bibliography

1 **Levine, I. N.**, Quantum Chemistry, Ira N. Levine, Editorial AC, 1977.

2 **Eisberg, R. M.**, Fundamentals of Modern Physics, Editorial Limusa, 1974

3 **Castellan, G. W.**, Fisicoquímica, Inter-American Educational Fund, 1976

4 **Levich**, Mecánica cuántica, Editorial Reverté, 1974

5 **Hobart, H. W.**, Métodos instrumentales de análisis, C.E.C.S.A., 1971

6 **Sanchez del Río,** Introducción a la teoría del átomo, Alhambra Editorial, 1977.

7 **Ronald Reese.**, Física Universitaria, Volumen 2, Edit. Thomson,2002.

8 **Burke, J.**, Física: La Naturaleza de las cosas, volumen 2, Edit. Thomson Paraninfo, 2001.

9 Hoffmann, **J.**, Matemática 5ª, Sphinx, Caracas, 1998

For a more complete bibliography see the one suggested in the text of Levine, I. N., Quantum Chemistry, Editorial AC, 1977.

www.ingramcontent.com/pod-product-compliance
Lightning Source LLC
Chambersburg PA
CBHW040313220526
45473CB00009B/2423